JOURNEY
to the EDGE
of REASON

ALSO BY STEPHEN BUDIANSKY

BIOGRAPHY
Oliver Wendell Holmes: A Life in War, Law, and Ideas
Mad Music: Charles Ives, the Nostalgic Rebel

HISTORY
Code Warriors
Blackett's War
Perilous Fight
The Bloody Shirt
Her Majesty's Spymaster
Air Power
Battle of Wits

NATURAL HISTORY
The Character of Cats
The Truth About Dogs
The Nature of Horses
If a Lion Could Talk
Nature's Keepers
The Covenant of the Wild

FICTION
Murder, by the Book

FOR CHILDREN
The World According to Horses

JOURNEY
to the EDGE
of REASON

The Life of Kurt Gödel

Stephen Budiansky

W. W. NORTON & COMPANY
Independent Publishers Since 1923

For information about permission to reproduce selections from this book, write
to Permissions, W. W. Norton & Company, Inc.,
500 Fifth Avenue, New York, NY 10110

For information about special discounts for bulk purchases, please contact
W. W. Norton Special Sales at specialsales@wwnorton.com or 800-233-4830

Manufacturing by LSC Communications, Harrisonburg
Production manager: Anna Oler

Library of Congress Cataloging-in-Publication Data

Names: Budiansky, Stephen, author.
Title: Journey to the edge of reason : the life of Kurt Gödel / Stephen Budiansky.
Description: First edition. | New York : W.W. Norton & Company, [2021] |
Includes bibliographical references and index.
Identifiers: LCCN 2020054854 | ISBN 9781324005445 (hardcover) |
ISBN 9781324005452 (epub)
Subjects: LCSH: Gödel, Kurt. | Logicians—United States—Biography. |
Logicians—Austria—Biography. | Mathematician—United States—Biography. |
Mathematician—Austria—Biography. | Gödel's theorem. |
Mathematics—Philosophy. | Logic, Symbolic and mathematical.
Classification: LCC QA29.G58 B83 2021 | DDC 510.92 [B]—dc23
LC record available at https://lccn.loc.gov/2020054854

W. W. Norton & Company, Inc., 500 Fifth Avenue, New York, N.Y. 10110
www.wwnorton.com

W. W. Norton & Company Ltd., 15 Carlisle Street, London W1D 3BS

1 2 3 4 5 6 7 8 9 0

CONTENTS

LIST OF MAPS AND
ILLUSTRATIONS

JOURNEY
to the EDGE
of REASON

At the Institute for Advanced Study, 1956

PROLOGUE

MARCH 1970. The psychiatrist moved his pen swiftly across the yellow sheets of lined notebook paper, recording facts, strange and mundane, about his new patient. Einstein had called him "the greatest logician since Aristotle," and even in Princeton, that town with more Nobel Prize winners than traffic lights, his otherworldly genius had stood out. The work he had done forty years earlier, at age twenty-four, had brought fame and recognition from around the world—"the most significant mathematical truth of the century," a staggeringly brilliant and paradoxical proof that no formal mathematical system will ever capture every mathematical truth within its own bounds.

But now he was tormented by demons, of failure and persecution. The psychiatrist wrote,

Kurt Gödel 64. Married 32 years. Adele 70. No children. Wife had been married once before.

Thought he came for evaluation of mental competency—which I denied—to "help him" if could. — Came at insistence of brother & wife.

Belief that he hasn't achieved goals that he set out for himself— hence a "failure"—therefore other people, particularly the Institute, will also regard him as a failure & try to get rid of him.

— Believes he has been declared incompetent & that one day they will realize he is free & take him away as being too dangerous.

Fear of destitution, loss of position at Institute because hasn't done anything for past year—has done hardly anything for 35 years—4–5 uninteresting papers. — Took on big subjects, may not have been talented enough. — Usually works on his own, in way & fields that are opposed to current stream. — Possible that he feels guilty about not being productive & achieve similar acclaim as he did as a young man.[1]

It was unseasonably warm for Princeton, a bright sun blazing through clear sky to push the temperature to a summerlike 71 degrees, when Gödel arrived for his first session with Dr. Philip Erlich. But inside the psychiatrist's office, a quiet brick house on Nassau Street that had stood since ten years before the American Revolution, he kept his overcoat on, complaining of the cold. Sometimes he would show up wearing one or even two sweaters as well, an odd counterpoint to the Old World formality that his dress and manner otherwise proclaimed: well-cut suit, sharp creases carefully pressed in trousers, grey hair

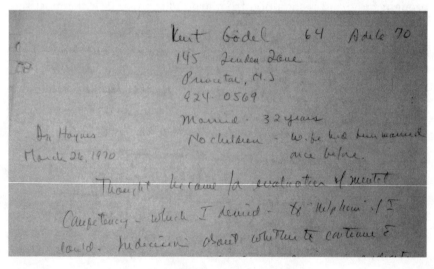

Dr. Philip Erlich's case notes, 1970

meticulously combed back from the forehead and distinguished with a streak of darker color in the very center that had oddly resisted the forces of age, scholar's large round spectacles, precise speaking voice whose clear and resonant cadences summoned up a picture of a more imposing and vigorous man than his perpetually underweight, five-foot-seven-inch frame suggested.[2]

He came twice a week—only, he said, because he feared his wife's rages if he failed to keep his appointments. In desperation she had summoned his brother from Vienna to help deal with the situation when things began to fall apart at the beginning of the year. Rudolf had arrived the first week of April but Gödel immediately had a fight with him, too.

Brought out delusional ideas. — Brother is the evil person behind plot to destroy him—because he wants to take his wife, house & position at Institute. — He also feels brother mishandled situation by becoming angry w/him, instead of remaining calm. I defended brother—motivated by good intentions, no desire to harm, called in by wife. — I emphasized need for firm action & insistence that p[atien]t see me.

Freud and his theories, the patient riposted, were manifestations of the same materialism that he himself completely rejected in his own work in logic and philosophy. The mind was based on far fewer physical foundations, and far more spiritual influences, than the twentieth century wanted to believe. The medieval thinkers had been right to look upon mental illness as a "spiritual infestation." Eventually the truth will be discovered—even though science is headed in a materialistic direction for the foreseeable future.

Once, he told the psychiatrist that he did not consider his consultations as therapy, but just as talking to a friend. All of his own friends were gone. During their years together in Princeton in the 1940s and early 1950s, he had been by far Einstein's closest companion. Einstein said his own work by then did not amount to much, but he came into the office "just to have the privilege of being permitted to walk home with Kurt Gödel."[3] They were an established Princeton sight, the almost comically mismatched pair making their way home each afternoon across the sweeping lawn of the Institute grounds, opposites in almost every personal trait. Einstein with his trademark unruly hair, baggy sweater, and suspenders—looking "like a good old grandpapa, which there is no reason to object to," Gödel had once said to his mother, defending his friend when she had made a wondering remark about Einstein's unkempt and "unaesthetic" appearance in a photo Gödel

Walking with Einstein in Princeton

had sent.[4] And Gödel solemn, serious, rail thin, immaculately dressed on the hottest summer's day in white linen suit and snappy fedora. Yet each day they walked and conversed, animatedly in German, on politics, physics, philosophy, life.

But Einstein had been dead for fifteen years. And then, the patient insisted, his one other old and close Princeton colleague, the genial and brilliant economist Oskar Morgenstern, whom he had known ever since the old days in Vienna, had dropped him, abandoned him for no reason. "I lost my best friend," he piteously said. Now the Institute for Advanced Study, the academic Elysium that had looked after him with paternalistic solicitude for three decades, was abandoning him, too. He was certainly about to be fired. Or perhaps had already been fired, the decision having been made in secret, and kept from him.

To all of the doctor's attempts to reason with him, the patient replied with the iron grip of logic:

> *Still is convinced of reality of his ideas, & to challenge them would be admission of insanity which would put into question the validity of his life's work. — Either I accepted him as objective & capable of reporting accurately & this led logically to the conclusion; or else spirits could be involved in terms of deceiving him.*

But surely, the psychiatrist objected, it would be false for the Institute's director to have just written you a letter affirming your status as a tenured professor with a guaranteed pension when you retire—"a permanent statement of the Institute's position," the director had emphasized—if he was going to fire you.[5]

Or, look at Einstein, he urged: Like you he did his great work as a young man, and he didn't become depressed.

Or once, in a counsel of complete despair: Try having a glass of sherry before meals.

He tried challenging his patient. You need a villain; sometimes the doctors fill that role, sometimes your brother. You suffer from a secret desire for incarceration from guilt over your perceived failings. You suffer from grossly inflated ego due to your early triumph and the honors showered on you as a young man.

The patient ridiculed all of these suggestions. He had never sought fame, he said; he had put that all behind him years ago. His only motivation throughout his life had been a desire for financial security and an interest in the work itself. But now he could not seem to do anything:

> *Listed the various distractions that have interfered w/his progress w/his philosophical work. Marriage, bookkeeping, his health problems mental & physical, wife's health, Institute duties, occasionally working on problems in pure mathematical logic, hobbies— reading history etc. — In his own case he feels that he becomes so involved w/preliminaries he never gets to the main body.*

Even his great work, the Incompleteness Theorem, was no longer a consolation. All of his contributions, he sadly observed, were of a negative kind—proving that something *cannot* be done, not what *can* be done.

For a while he had started to seem better, showing flashes of his old gently humorous self, regaining his lost weight, even resuming some of his work and appearing at his office to the surprise of his colleagues, and he stopped coming to see Dr. Erlich after eleven months of their twice-weekly and then weekly sessions.

But in 1976, everything began to fall apart again. He was refusing urgently needed prostate surgery, was not eating, was preoccupied with his physical condition, and paranoia and self-loathing again consumed him. Of his final few desperate visits to Dr. Erlich in the year before his death, the psychiatrist wrote:

> *Situation deteriorating. — Digging his hole deeper. — Believes he was fired by faculty 1 year ago. — Strong self hatred & fear of punishment. Also self torment over minor things. — Blurting out to people his errors that are none of their business & only put him in a bad light. — Very difficult, willful man.*

A few months later, in January 1978, he was dead. His weight at death was sixty-five pounds.

Dr. Erlich interpreted his patient's refusal to eat in his last days as the final suicidal act of a man tormented by unresolved guilt. But the attending physician at Princeton Hospital saw it differently: "More apathy and resignation than an active volitional suicidal effort."[6]

In the end, all he had left were negative decisions.

· 1 ·

Dreams of an Empire

THE GOLDEN AGE OF SECURITY

Few worlds could have been farther apart than the tormented darkness of Kurt Gödel's final years and the golden security of his youth, in what in 1906 still seemed the eternal Habsburg Empire. From the attic skylight of the pleasant villa that his father had built at the base of the wooded slopes leading up to Brünn's medieval Spielberg Castle, he and his brother could look out on clear days and glimpse in the distance the smoke of locomotives heading along the railroad line to Vienna, just seventy miles to the south.[1]

Administratively a part of the predominantly Czech-speaking land of Moravia, the city of Brünn was as firmly a piece of the great Austrian state as any of its other ancient bastions that had been united for centuries under the Habsburg dynasty. The rail lines that spanned the empire's breadth of nearly a thousand miles, from the Adriatic Sea in the southwest to the Russian frontier in the east, linked its great centers of Trieste, Prague, Budapest, Cracow, Czernowitz, Graz, Lemberg, and hundreds of smaller towns and villages in between to the imperial capital of Vienna at its very heart. Those bands of steel were but a recent manifestation of the Austrian Empire's age-old stabilizing stronghold in Central Europe.

Across the realm, every large to moderate-sized city sought to imi-

The rail line to Brünn; Spielberg Castle and Cathedral of
Saints Peter and Paul in the distance

tate the architecture of the capital. A broad ring street; a municipal
theater or opera house, constructed in the same ponderous and cacoph-
onous blend of neo-baroque, neo-classical, and neo-Renaissance architectural styles as its prototypes on Vienna's Ringstraße; rows of stolidly
bourgeois apartment blocks behind aristocratic facades of stucco ornament. Pulling in to even the smallest town of the empire, an Austrian
would be greeted by reassuringly familiar sights and sounds: the same
slightly overweight middle-aged stationmaster, paunch stuffed into his
inoffensive dark blue uniform, saluting each train upon its arrival as a
kind of "military blessing," in the words of the Austrian novelist Joseph
Roth; the same small bell, carried on a black belt across his chest, ringing the identical treble note from the Adriatic to the Russian frontier
to signal the train's imminent departure. Every station, painted bright
yellow, was watched over by the ubiquitous black double-headed eagle

of the emperor's seal affixed to the wall. The *Fiaker* drivers waiting out front for fares made the same jokes, gesticulated the same gestures, addressed their customers in the de rigueur Austrian manner by which everyone was promoted to the rank above the one he actually held: every businessman was *Herr Direktor*, every army major was *Herr General*, every university student was *Herr Doktor*.[2] Even the most far-flung provincial coffeehouse could have been a duplicate of its Viennese original, down to the smoke-stained walls, chessboards and domino sets, cake trolleys and blue-aproned waitresses, trays of kaiser rolls and poppy seed loaves, and identical buxom blonde cashier keeping a sharp eye over business.[3]

A year before Roth drank himself to death in exile in Paris in 1939, one of the countless victims of what the historian George Berkley called "the most unrequited love affair in urban history"—the tragic devotion of Vienna's 300,000 Jews to the country that had given them and 2 million of their kinsmen an unparalleled haven and hope only to see it ground to ashes—he described this mystical sense of unity, which had managed to defy the fragmenting forces of nineteenth-century nation-alism and hate.[4] Looking back only after it is all gone, the narrator of Roth's final novel, *The Emperor's Tomb*, reflects that

> only much later, then, was I to realize that even landscapes, fields, nations, races, huts and coffee houses of the most widely differing sorts are bound to submit to the perfectly natural dominion of a powerful force with the ability to bring near what is remote, to domesticate what is strange and to unite what seems to be trying to fly apart. I speak of the misunderstood and also misused power of the old Monarchy which worked in such a way that I was just as much at home in Zlotograd as I was in Sipolje or Vienna.[5]

And so it was in Gödel's Brünn, from its ring street, to its Stadtthe-ater where touring companies from Vienna staged the latest plays, to the Gymnasium that taught the universal curriculum of the empire to high school students planning to continue their studies at the University of Vienna or Prague or Czernowitz, to the Café Schopp on the Freiheitsplatz where Gödel's mother spent leisurely afternoons with friends: the changes about to sweep it all away seemed unthinkable in the years before the First World War.

Those years, recalled the writer Stefan Zweig, another contemporary of Gödel's, and like Roth another Austrian Jew who managed to escape Hitler only to die by his own hand in heartbroken exile, were the "Golden Age of Security." The currency, the Austrian crown, circulated in "shiny gold coins, thus vouching for its own immutability." Parents set aside savings for their children as babes in the cradle, knowing the money would be there for their future. A civil servant had only to consult the calendar to know when he would be promoted and when he could retire on an assured pension. "Everything in this wide domain was firmly established, immovably in its place, with the old Emperor at the top of the pyramid, and if he were to die the Austrians all knew (or thought they knew) that another would take his place, and nothing in the well-calculated order of things would change. Anything radical or violent seemed impossible in such an age of reason."[6]

Brünn's Stadttheater, 1905

Stability was the personal virtue that stood above all others, and that faith was rewarded by a society that could be counted on to return it in full measure. "The irrational, the passionate and the chaotic," says one historian of the era of Austrian certainty, "were to be avoided at all costs."[7]

The "Austrian Idea," a phrase coined by the Austrian poet and librettist Hugo von Hofmannsthal in 1917 just as it vanished forever in the paroxysm of the Great War, was one of "reconciliation, synthesis, bridging of differences," of embracing and carrying across the border-lands of race and language universal values of European civilization and culture. "Austria is neither a state, a home, nor a nation," says the mad brother of one of Roth's characters, a Polish count from the Austrian territory of Galicia. "It is a religion." It is thus not a multi-national state, but a *supra*-national one, he explains: "the only supra-nation which has existed in this world." ("As a private person my brother is as mad as a hatter," says the count, "but where politics is concerned he is second to none"—one of Roth's many wry allusions to the schizophrenic realities of Austrian life.)[8]

It was an idea that was first forced upon, then fully embraced by the Austrian monarchy. Standing astride East and West, Vienna had for centuries been a crossroads and an enforced melting pot. A school-master from the Palatinate visiting the city in 1548 claimed to have identified a dozen and a half languages being spoken in the merchant quarter, from Polish, Hungarian, Croat, Czech, and Italian to Turkish, Greek, Hebrew, and Arabic.[9] Frederick III, the first Habsburg to be crowned Holy Roman Emperor, in 1452, formulated Austria's *mission civilatrice* in his famous and immodest acronym *AEIOU*, which he plastered on everything from cathedrals to his tableware. In German it stood for *Alles Erdreich ist Österreich untertan*: "It is for Austria to rule the whole world." For the next two centuries Austria came very close to doing just that, at least the "whole world" of Christian Europe, in which the Habsburg monarchy reigned over Spain, the Nether-lands, most of Italy, and parts of France, besides its heartland in Cen-tral Europe.

THE GOOD HABSBURG

By the Age of Enlightenment of the late eighteenth century, Austria had lost Spain but had acquired Hungary and Transylvania to the east, along with the emperor who would still be recalled by Austrians two centuries later as "the good Habsburg" (or, more mordantly, "the only Habsburg who was not an idiot"). In his zeal to modernize the administration of the state during his reign from 1765 to 1790, the reform-minded Joseph II did much to make the Austrian Idea reality. He professionalized the legal system, curtailing the arbitrary power of local nobility and establishing equality before the law. He redrew and rationalized the boundaries of administrative districts, abolished serfdom, and allowed the peasants to purchase hereditary tenure over the lands they tilled. He shuttered monasteries and put the proceeds to a better use building schools. He founded in Vienna the largest and most modern hospital in the world and the country's first humanely constructed institution for the mentally ill (promptly dubbed, with the usual, casual Viennese cruelty, the *Narrenturm*, or Tower of Fools), which today still stands as the home of a morbid collection of pathological anatomical specimens. In between he found time to be the patron of Mozart, during the composer's last brilliant years in Vienna.[10]

Joseph II dreamed of an ideal *Beamtenstaat*, a perfect administrative order in which a secular priesthood of dedicated, loyal, and efficient public servants, motivated only by "a burning enthusiasm for the good of the state," as he envisioned it, would spread rationality and order throughout his empire.[11] He issued hundreds of decrees, regulating the width of streets, the sanitation of towns, the price of funerals, the management of forests, the inspection of bridges and mines, the supervision of food markets, the establishment of brick factories and coffeehouses, the care of the aged and indigent, and the siting, size, and furnishing of schoolhouses, down to the precise location of the blackboard in each classroom. By 1800 there was hardly a public official to be found who lacked a university degree, and the bureaucracy's growing cultural identification with the imperial center did much to ensure that Vienna set the tone for the entire empire.[12] No less an authority than Adolf Hitler, who nurtured a provincial-born Austrian's enduring resentment of the sophisticated metropolis, testified how difficult it was even after he had subsumed his old homeland into the Reich to depose Vienna from its mystic place as the old Habsburg capital. "It was a tremendous task," the Führer was still fuming in April 1942, when he also presumably had other things to think about, "to break Vienna's cultural preponderance" over the Austrian territories he had annexed in the 1938 *Anschluss*.[13]

Joseph believed that in return for good governance, the grateful people of his sprawling empire would rationally choose to redirect family, caste, clan, and national loyalties in favor of the *Vaterland* that provided them security, and—shades of the words being written at that very moment six thousand miles away, in America's Declaration of Independence—the gift of "happiness."[14]

At the turn of the twentieth century Austria shared with only one other nation, the United States of America, the distinction of being a country that was more an idea than a national tribe. Though German remained the language of administration and lingua franca throughout the Austrian half of the Austro-Hungarian Empire, nearly all Austrians, and certainly all members of its ruling class, had long regarded themselves as anything *but* German. Native German speakers had always been and remained a distinct minority within the empire. By the end of the first decade of the twentieth century Austria's population of 28.6 million included 10 million Germans, 6.4 million Czechoslovaks, 5 million Poles, 3.5 million Ruthenians (i.e., Ukrainians), 1.25 million Slovenians, 780,000 Serbo-Croats, 770,000 Italians, 275,000 Romanians, plus some half a million foreigners. Hungary's 20.9 million included 10 million Hungarians, 2.9 million Romanians, 2.9 million Serbo-Croats, 2 million Germans, 2 million Slovaks, and 470,000 Ruthenians.[15]

The German that *was* spoken by Austrians throughout the empire had long acquired distinctive features that marked their enduring dif-

Austro-Hungarian Monarchy, 1906

ferences from German Germans, just as Austria's solid Catholic faith set them apart from predominantly Protestant northern Germany. Austrian German, besides possessing its own unique term of abuse for its neighbors to the north (*Piefke*, roughly translatable to English as "Kraut"), is filled with words borrowed from across its polyglot empire that leave speakers of "high" German scratching their heads: *Tollpatsch* (Hungarian *talpus*, clumsy oaf), *Jause* (Slovene *južina*, afternoon snack), *Feschak* (Czech *fešák*, handsome guy), *sekkant* (Italian *seccante*, annoying), and a raft of Yiddish words, among them *meshuggah* (crazy), *Ganif* (thief), and *Mischpoke* (family).

Genuine Viennese slang, the unique dialect known as *Wienerisch*, is almost incomprehensible to outsiders, but much of its characteristics infused even the "Schönbrunn German" spoken by the emperor and his entourage, the senior ranks of the civil service and the army, and the upper bourgeoisie; it was, as one historian describes it, "a soft idiom presumed to meet that of the people half-way."[16]

Like its words, its nobility, its army, and its officialdom, Austria's distinctive cuisine was also a sum of its multifarious parts, the German elite having contributed in fact almost nothing in the way of original sources to the popular dishes of Vienna's celebrated coffeehouses, hotels, and restaurants. The coffee was Ottoman, the apple strudel Hungarian, the pastries and apricot dumplings Czech. Even that most famous of all Viennese dishes, the *Wiener schnitzel*, was nothing but an Italian *costoletta alla milanese* cloaked in a German name.

Though it was not until 1867 that full equality in political rights was granted to all of the empire's subjects, Jews included, the idea of the state as the overarching protector of its multiple constituent nationalities was so much a part of Austrian consciousness that even during the 1848 revolutions that swept Europe, Austria's rising Slavic nationalist movements saw themselves as "caretakers of the true Habsburg imperial idea," in the words of the historian Pieter M. Judson. They sought to reshape rather than overthrow the empire. It was during that year of revolution that a Czech patriot wrote the most famous, if usually misquoted, line about Austria. František Palacký's words are often reduced to little more than a flip witticism: "If the Austrian Empire did not exist, it would have to be invented." But what he actually said was something much more profound, and sincere, about the unique role the empire played as an anchor of stability, for both Europe and its subject peoples: "Had the Austrian state not existed for ages, it would assuredly have been in the interests of Europe, and indeed of *humanitas*, to endeavor to create it as soon as possible."[17]

The Austrian state, against all odds and all of the tendencies of the nineteenth-century *Zeitgeist*, managed to make reactionaries even out of revolutionaries.

"THE HALF-TRODDEN PATH AND THE HALF-COMPLETED ACT"

The stability of nineteenth-century Austria was embodied in the person of its long-living, mutton-chop-whiskered, and in the end much beloved emperor Franz Joseph I, who reigned from 1848 to 1916. He was in some ways a comic-opera royal, addicted to uniforms and pomp, hunting and horses, rising every morning at five o'clock in the cavernous, uncomfortable, and tastelessly but elaborately gilt- and red-plush-decorated rooms of the old Hofburg palace, to prepare for a rigidly prescribed daily routine of royal audiences, military inspections, and state ceremonies. Every pupil in the land learned by heart the emperor's formal title, redolent of grandeur, history, and the dead weight of anachronism:

> His Imperial and Royal Apostolic Majesty, By the
> Grace of God Emperor of Austria, King of Hungary
> and Bohemia, Dalmatia, Croatia, Slavonia, Galicia,
> Lodomeria and Illyria; King of Jerusalem, etc.; Archduke
> of Austria; Grand Duke of Tuscany and Cracow; Duke
> of Lorraine, Salzburg, Styria, Carinthia, Carniola and
> Bukovina; Grand Prince of Transylvania, Margrave of
> Moravia; Duke of Upper and Lower Silesia, of Modena,
> Parma, Piacenza and Guastalla, of Auschwitz and

Zator, of Teschen, Friaul, Ragusa and Zara; Princely
Count of Habsburg and Tyrol, of Kyburg, Gorizia
and Gradisca; Prince of Trent and Brixen; Margrave
of Upper and Lower Lusatia and in Istria; Count
of Hohenems, Feldkirch, Bregenz, Sonnenberg etc.;
Lord of Trieste, of Cattaro and on the Windic March;
Grand Voivode of the Voivodeship of Serbia etc., etc.

The whole impressive if mendacious claim was publicly proclaimed in the grand ceremony by which Austria's emperors, albeit only posthumously, made a magnificent show of Christian humility. When the funeral procession arrived at the small Capuchin church that housed the royal crypt, down a tiny winding side street from the grand Hofburg palace, the Grand Chamberlain would beat upon the door and announce, "I am . . ." and proceed to recite the emperor's full title. The Capuchin friar would reply, "I know him not." Refused again as "His Majesty the Emperor and King," the Chamberlain would upon the third knock at the door identify the man seeking entry to his final resting place simply as, "A poor mortal, and a sinner," whereupon the doors would at last swing open.

Though Franz Joseph did much to make Austria a modern, prosperous, and superbly well-educated state, he always did so hesitantly, reluctantly, with one eye to the past. The nineteenth-century Viennese playwright Franz Grillparzer spoke of "the half-trodden path and half-completed act" that was the curse of the Habsburgs. Joseph II's successor Franz I of Austria, who occupied the throne from 1792 to 1835, opposed change as a matter of principle, regarding any alteration to the status quo as a threat to the empire itself. "My realm," he remarked, "resembles a worm-eaten house. If one part is removed one cannot tell how much will fall." He instructed teachers to stay away from "new ideas" altogether: "I do not need scholars, but rather honest citizens."[18]

For his part, Franz Joseph temporized and backtracked on every reform touching the great questions of democracy, nationalism, and the power of the Catholic Church over civil affairs. Even in his personal life he conceded as little as possible to the modern age. Throughout his reign the Hofburg remained lit by kerosene lamps (the emperor said electric lights irritated his eyes) and the heat came from immense ceramic stoves in the corner of each chamber. The primitive toilet facilities of the palace so irritated his daughter-in-law Stephanie that the princess finally had two bathrooms installed at her own expense. Court manner, dress, and table service were dictated by the Spanish Court Ceremonial of three centuries earlier. The story was told that on his deathbed, Franz Joseph was outraged by the attire of his hastily summoned physician. "Go home and dress correctly," the dying emperor commanded.[19]

Joseph II's machinery of rational government had continued to grow under his successors, but as often as a tool of repression and stasis than of enlightenment. It was said that Franz Joseph ruled with four armies: one that marched (the military), one that knelt (the church), one that sat (the bureaucracy), and one that crawled (the secret police and its thousands of informants). The bureaucracy, which numbered 100,000 in the 1860s, grew to 300,000 by 1900. It was more feared than respected, able to make life miserable through formal adherence to pettifogging rules, as it became less a rationally designed machine than a crazy jury-rigged superstructure that attempted to reconcile the irreconcilable contradictions of the empire.[20] In 1867, Franz Joseph acceded to a complex compromise aimed at defusing the nationalist powder keg just enough to let him hold onto the eastern half of his realm. Hungary was granted its own parliament, in which Franz Joseph sat as King of Hungary, not Emperor of Austria. The result was not two but three separate bureaucracies: one for the kingdom of Hungary, one for the empire of Austria, and a third for joint affairs of war, foreign affairs, and finance. "K. und k.," the ubiquitous title of state institutions in Franz Joseph's era, stood for *kaiserlich und königlich*, "Imperial and Royal," reflecting his dual role. The Austrian half of the realm was left with no real name at all; its official designation was only "The Kingdom and Provinces Represented in the Reichsrat," while the unofficial term, used by the bureaucrats, was the even less euphonious "Cisleithania," referring to the river Leitha that separated Austria and Hungary. The Austrian writer Robert Musil, in his great 1930 novel set in the time of the Austro-Hungarian monarchy, *The Man Without Qualities*, called the whole mess Kakania—a play on "k. und k." that of course also literally meant "Shitland."[21]

The army, in contrast to the bureaucracy, was more respected than feared. Throughout Franz Joseph's reign it lost war after war, bits of the empire dropping off each time. But it could always put on a good parade. Stefan Zweig remarked that the army had better bandmasters than generals; and its colorful uniforms, which Sigmund Freud likened to a parakeet's plumage, brilliant green and blue tunics over snow-white trousers, brightened every formal occasion. Despite its unbroken series of battlefield humiliations, writes the historian William M. Johnston, "no army in Europe was more popular."[22] One of the great set-pieces of Joseph Roth's masterpiece *The Radetzky March* portrays the spectacle of the annual Corpus Christi procession which could make even the most cynical Viennese skeptic, at least for a moment, believe in the Austrian dream:

> The blood-red fezzes on the heads of the azure Bosnians burned in the sun like tiny bonfires lit by Islam in honor of His Apostolic Majesty. In black lacquered carriages sat the gold-decked Knights of the Golden Fleece and the black-clad red-cheeked municipal counselors. After them, sweeping like the majestic tempests that rein in their passion near the Kaiser, came the horsehair busbies of the bodyguard infantry. Finally, heralded by the blare of the beating to arms, came the Imperial and Royal anthem of the earthly but nevertheless Apostolic Army cherubs—"God preserve him, God protect him" . . . The loud fanfares resounded, the voices of cheerful heralds: Clear the way! Clear the way! The old Kaiser's coming!
>
> And the Kaiser came; eight radiant-white horses drew his carriage. And on the white horses rode the footmen in black gold-embroidered coats and white periwigs. They looked like gods and yet they were merely servants of demigods. On each side of the carriage stood two Hungarian bodyguards with a black-and-yellow panther skin over one shoulder. They recalled the sentries on the walls of Jerusalem, the holy city, and Kaiser Franz Joseph was its

king. The Emperor wore the snow-white tunic well known from all the portraits in the monarchy. . . . The Kaiser smiled in all directions. . . . The bells tolled from St. Stephen's Cathedral, the salutes of the Roman Church, presented to the ruler of the Holy Roman Empire of the German Nation. The old Kaiser stepped from the carriage, showing the elastic gait praised by all newspapers, and entered the church like any normal man; he walked into the church, the Holy Roman Emperor of the German Nation, immersed in the tolling of bells.

"No lieutenant in the Imperial and Royal Army," concluded Roth— by which Roth really meant no subject of the Austro-Hungarian Empire—"could have watched this ceremony apathetically."[23] Franz Joseph, with his proclamations which began with the words, "To My Peoples," was looked upon by those peoples, in a very real sense, as the personal guarantor of their equal rights in his strange polyglot realm, and so the guarantor of the empire itself. His presence in the national consciousness was a blend of the ethereal and homey. The cautionary political catchphrase of Franz Joseph's era encapsulated it perfectly: "Surely you can't do that to the old man."[24]

THE MECCA OF SCIENCE AND MEDICINE

The industrial revolution, the stirrings of liberal democracy, the new age of discovery in science and medicine, and the rise of a burgeoning middle class, all hallmarks of nineteenth-century European civilization, had been so long in coming to Austria that when they did finally arrive the effect was to inject a feeling of exuberance into life unmatched anywhere on the Continent. The construction of Vienna's fabled Ringstraße, itself made possible only by the imperial city's backwardness that had preserved a large, concentrated undeveloped swath

Opposite, top: Vienna's medieval walls and glacis
Opposite, bottom: The Ringstraße nearing completion: (l. to r.) Parliament, Rathaus, University (with spires of Votiv Church behind), Burgtheater

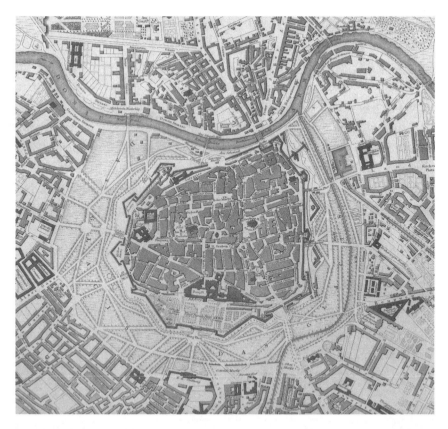

of land at its very center, was the most visual manifestation of the new energy that galvanized the life of the old empire.

Franz Joseph's order to clear away the city's ancient fortification walls and gates along with the sloping, half-kilometer-wide glacis in front (constructed to give troops defending the Hofburg a free field of fire) turned a medieval bastion into a beautiful modern city of broad tree-canopied boulevards, interconnecting parks, and monumental new public buildings. Intended to declare Vienna's unquestioned place as the capital of a great empire, the new buildings were designed in what would be generously termed the "historicist" style, part ancient, part modern, and part Disney, each declaring its place in the pantheon of civilization: The parliament a Greek temple, the art museum a baroque palace, the university a Renaissance court, the Votiv Church a Gothic cathedral. They were all something of a pastiche, the new opera house most of all, which seemed to incorporate every conceivable style in a single building. A derisive ditty in the local Viennese dialect satirized *der Ringstraßenstil* and its two famous architects:

> *Der Sicardsburg und van der Nüll,*
> *Haben beide keinen Stüll.*
> *Griechisch, gotisch, Renaissance,*
> *Das ist ihnen alles aans!*

> *That Sicardsburg and van der Nüll,*
> *Have no style that you can name.*
> *Gothic, Greek, and Renaissance,*
> *To them they're all the same!*

But as an example of successful urban planning the Ringstraße was, and remains, superb, the most visually striking of any city of the time, full of life and vitality, culture and commerce, its brightly lit coffeehouses, elegant hotels, grand town palaces of the Ephrussis, Rothschilds, Wittgensteins, and the other great families whose fortunes grew with Austria's plunge into commerce and industry, all woven into Vienna's legendary charm. The channeling of the Danube to control flooding that had plagued the city for centuries, the completion of aqueducts from distant Alpine springs that supplied the famously pure water of Vienna's municipal water system, and other civic improvements justified on the basis of *Verschönerung des Stadtbildes*—the beautification of the city's image—added to the city's livability and sense of rapid progress.[25]

By 1900 Austria's economy was growing faster than Great Britain's, as a wave of prosperity from trade and the manufacture of textiles, steel, glass, machinery, musical instruments, and other goods lifted the middle class.[26] Gödel's immediate forebears on both his mother's and his father's side had been beneficiaries of this rising tide of prosperity and security. In two generations, they had gone from leatherworkers, hand weavers, and bookbinders to managers of the booming textile mills established in Brünn in the mid-nineteenth century.

Austria's rise as a center of culture, intellect, and science was equally striking; that too had been heightened by its sudden appearance against an unassuming backdrop. Madame de Staël, a cultured Frenchwoman who visited Austria in 1808, found little to write home about when it came to science and letters. "It is such a calm country," she reported, "a country where everyday comforts are so tranquilly secured to all classes of citizens, that here one does not think much about intellectual pleasures."[27]

By the end of the century, Austria not only had one of the largest and most comprehensive educational systems in Europe, but had become a recognized world leader in medicine, physics, philosophy, and mathematics. The German-born Rudolf Virchow, a pioneering nineteenth-century cell biologist and pathologist, dubbed the Vienna of this period "the Mecca of Medicine." During the second half of the nineteenth century, Viennese doctors—a good many of whom came from Bohemia, Moravia, or other lands of the Austrian Empire—contributed an unbroken string of medical firsts. Adolf Lorenz invented a noninvasive surgical technique to correct club footedness; Ferdinand von Arlt discovered the cause of nearsightedness and Eduard Jäger von Jaxtthal developed the eye chart to standardize prescriptions for glasses; Vincenz von Kern revolutionized the treatment of wounds and Theodor Billroth pioneered the use of ether and chloroform for anesthesia; anatomist Carl von Rokitansky, carrying out tens of thousands of autopsies, laid the foundation of modern pathological diagnosis; and most famously of all, Sigmund Freud, born in Freiberg in the far northeastern reaches of Moravia, had already begun to expound his revolutionary ideas of the human psyche. The reputation of the University of Vienna's medical school drew students from across Europe, and increasingly from America, too.[28]

In the natural sciences, Austria produced a remarkable number of talented theorists and experimentalists. The electrical genius Nikola Tesla, from Croatia, studied in Karlovac at one of the rigorous German-language high schools, the Gymnasiums, established throughout the Austrian Empire. Carl Menger, creator of the marginal utility theory, which revolutionized economics by demonstrating that prices are determined not by the costs of materials and labor but by demand for each additional unit of the product, was from Polish Galicia, and had studied in Prague, Vienna, and Cracow. The Viennese-born Ludwig von Boltzmann, one of the giants of physical chemistry, developed the theory of statistical mechanics, which relates the statistical behavior of atoms to physical properties of matter such as heat capacity of a metal or pressure of a gas. And from Gödel's hometown of Brünn came the geneticist monk Gregor Mendel and the experimental physicist Ernst Mach, whose name would be immortalized in the unit of the speed of sound in recognition of his exploration of supersonic dynamics—which led him along the way to the development of high-speed photography, which he used to capture extraordinary images of bullets in flight and their attendant shockwaves.

Though the University of Vienna or one of the great German universities like Göttingen or Berlin was where every Austrian professor hoped to wind up, most had to start at posts in Austria's far-flung provinces, which had the beneficial result of spreading new advances in the sciences throughout the empire. Gödel's dissertation adviser and mentor Hans Hahn was entirely typical in having spent six years at the Franz Joseph University in Czernowitz as his first professorial appointment.

Hungary was drawn along in the same vortex of intellectual excitement and scientific progress that enveloped the rest of the empire. An extraordinary constellation of the twentieth century's leading physicists and mathematicians were the product of its equally exceptional educational system at the turn of the century—John von Neumann, Edward Teller, Leo Szilard, Eugene Wigner, Theodor von Kármán, Paul Erdös, and George Pólya, among many others. All came from Hungary's Jewish middle class, all would flee Hitler's Europe, and many would end up working during the Second World War for the Manhattan Project, helping to ensure that America, and not Germany, would be the first to build the atomic bomb.

The educational reforms instituted in the era of ascendant liberal values in the last decades of the Austro-Hungarian Empire emphasized creative thinking and experimental curiosity over rote learning. Theodor von Kármán, whose permanent move to America in 1934 deprived the Reich of the greatest aerodynamicist of the age, was "von" because of the baronetcy conferred upon his father by Franz Joseph for his work modernizing the school system in the 1890s. "At no time did we memorize rules from a book," remembered von Kármán of the enlightened approach to learning his father had introduced. "Instead we sought to develop them ourselves." Leo Szilard, who discovered the nuclear chain reaction in the 1930s, recalled the Budapest of his youth as a society where economic security was taken for granted, and the highest value placed on intellectual achievement.[29]

The legendary comfort of Austrian society, its devotion to the *douceurs de la vie* that Madame de Staël had observed in 1808, had not been lost in the new atmosphere of intellectual excitement, but rather only added to the charm of the intellectual world of which Vienna could now boast. Enjoying a leisurely afternoon at any one of the dozens of its famous coffeehouses with their marble tabletops, ornate hanging lamps, racks of free newspapers to peruse while sipping a strong black coffee topped with a huge mound of rich whipped cream, always brought by a waiter on a silvery tray accompanied by a glass of cold water; or waltzing away an evening at one of the innumerable balls that filled the long pre-Lent season, or taking in the "charming inanities" of a Viennese operetta; or a weekend excursion to one of the rustic wine taverns just a short tram ride up to the foothills where the city's own vineyards began, which served each year's new wine along with a panoramic view of the city and Danube stretching out below—experiencing all of this, in a city now also renowned throughout the world for its unsurpassed art galleries, medical clinics, and scientific preeminence, a Viennese, as the historian William Johnston said, "might with equanimity regard his city as the pivot of the universe."[30]

LAND OF OPPORTUNITY

Though Gödel was not Jewish, it was no coincidence that growing up in the textile center of Brünn, and later studying at the great intellectual magnet of the University of Vienna, he found that almost all of his closest friends were Jews or of Jewish descent. No group had benefited more from economic liberalization and new educational opportunities. Friedrich Redlich, his father's employer and later partner in the textile mill where he worked as a manager, was born to Jewish parents and converted to Lutheranism as an adult; all of Gödel's best school-friends at the scientifically oriented Realgymnasium he attended in his secondary-school years in Brünn were Jewish; and in the brilliant circle of mathematically and philosophically inclined students at the university in which he quickly found himself, nearly all were Jewish, as were most of the professors who most deeply influenced him.

Barred for centuries from the universities, professions, and craft guilds, subjected to punitive taxation and legal restrictions that had kept Vienna's Jewish population to no more than a few thousand, Jews responded to Franz Joseph's 1867 decree granting complete religious freedom and equal civil rights in a burst of long-thwarted enthusiasm. The traditional Jewish zeal for education made itself felt almost instantly in their contributions to Austrian medicine, science, and literature. By the 1890s, although only 5 percent of the empire's population and 10 percent of Vienna's, Jews made up 40 percent of the students who attended the city's rigorously academic Gymnasiums, the prerequisite for entering a university. Approximately 30 percent of students at the University of Vienna were Jews; in the medical school, the figure was closer to 50 percent.[31]

The industrial revolution, regarded warily by traditional craftsmen and small merchants as a threat to their livelihood, likewise represented nothing but opportunity to Jews, who had nothing to lose and everything to gain by new economic freedom unconstrained by guilds, aristocrats, or national boundaries.[32] The great Jewish banking families made much of their early fortunes in these new areas of manufacturing and international trade: the Wittgensteins in steel; the Rothschilds in railroads; the Ephrussis in grain and oil. Austria's Jews were pioneers in the textile industry centered in Moravia and Lower Austria, and established Vienna's first department stores.

In all of these new endeavors Austria's Jews possessed a number of unique advantages, which quickly propelled them to the front ranks. Literacy rates even in the poorest, tiniest, and most traditional Jewish communities in the far eastern reaches of the empire approached 100 percent. Jewish boys from the earliest age acquired three or four languages: the Yiddish spoken at home; Hebrew, which they learned to read and write beginning at age three to study the Torah, followed by Aramaic for the Talmud; the local vernacular of Polish, Hungarian, Czech, Ukrainian, or Romanian; and often German as well, the official language of the empire's administration and top schools. The prestige of study that was deeply ingrained in Jewish tradition transferred itself readily to the secular world, along with a burning desire to catch up on what had so long been denied them. Teachers in the Gymnasiums across Austria and Germany noted the thirst of their Jewish students for knowledge, their diligence in their studies, and the close involvement of their parents, who "carefully monitor their progress" and continually impress upon them the importance of education.[33]

As in industry, Jewish scholars were more willing to take creative risks and to experiment in the sciences and arts: here too they had less to lose, as a result of having no vested interests in the status quo. Freud spoke of how that past marginalization became a strength in formulating his own groundbreaking ideas: "Because I was a Jew, I found myself free of many prejudices which restrict others in the use of the intellect: as a Jew, I was prepared to be in the opposition and to renounce agreement with the 'compact majority.' "[34]

The Jewish love for the written word infused Vienna's literature and theater with a new wave of exuberance in the latter decades of the nineteenth century as well. Karl Kraus, a Catholic convert from a Bohemian Jewish family, who enjoyed a huge and devoted following for his bitingly satirical journal *Die Fackel*, called himself a dweller "in the old house of language." With their reverence for German literature and culture, Jews dominated Austrian journalism and letters; more than 50 percent of the members of the journalists' society of Vienna circa 1900 were of Jewish ancestry, as were many if not most of the leading novelists, operetta librettists, and playwrights of fin-de-siècle Vienna.[35]

The sudden Jewish prominence in all of these fields appeared all the greater against a backdrop of conservatism and lassitude of Austrian Catholic society. Franz I's dislike of new ideas reflected a long and deep Catholic suspicion of too much education for its adherents, reinforced in Austria by the Counter-Reformation's fight against freethinking Protestant individualism. The philosopher Ludwig Wittgenstein would encounter these attitudes alive and well when he taught school in a small Austrian village in the 1920s and was constantly battling the parents who saw no good in filling their children's heads with ideas. Those attitudes extended to culture as well—notoriously expressed by the deputy in the Reichsrat from the Christian–Social party who contemptuously declared, "Culture is what one Yid cribs from the other."[36]

But the Jews of Vienna filled the theaters, devoured the latest novels, delighted in the wittily deft prose of the newspaper columnists, and generally reveled in their newfound place in a society where intellectual vitality, material prosperity, and culture combined. "We enjoyed the splendid city which was so elegantly beautiful," later wrote Max Graf, a Jewish music critic and friend of Freud's in Vienna, "and never thought that the light which shone over it could ever be that of a colorful sunset."[37]

LAND OF *SCHLAMPEREI*

With the benefit of hindsight, there were many signs of what was wrong just beneath that splendid surface. Beyond the Ringstraße, with its lively cafes, neat bourgeois apartment houses, and town palaces built on the nation's new industrial wealth, stretched miles of desperately overcrowded working-class neighborhoods unequal to housing a population that more than tripled from 1850 to 1900. Only a tiny percentage of single-family homes existed; most apartments were cramped, dark, ill-heated, and lacking basic sanitation. Even by 1910, 78 percent of apartment buildings had no indoor toilets and 93 percent no baths. The city's flophouses—where an impecunious Adolf Hitler stayed at times during his years in Vienna from 1908 to 1913, dreaming of becoming an artist—were notorious for their armies of voracious bedbugs, filthy communal living spaces, and rampant tuberculosis, known as the "Viennese disease." The average worker subsisted on a diet of little more than coffee, rolls, soup, and bread, with the occasional sausage and glass of beer.[38]

The pavements were so crowded with prostitutes offering their wares "in every price range and at every time of day," Stefan Zweig recalled, that "it really cost a man as little time and trouble to hire a woman for a quarter-of-an-hour, an hour, or a night as to buy a packet of cigarettes or a newspaper." Prostitution was a response not only to the desperation of women unable to afford their rent, but to the stultifying bourgeois society that left young men few sexual outlets. In a society that worshipped stability and security, men in their twenties were consigned to a kind of extended purgatory. A middle-aged man in fin-de-siècle Vienna affected not youth but dignity: graying beard, portly waistline, grave demeanor, and an unhurried gait whenever walking down a street or ascending a staircase. Early marriage was frowned upon; a man was expected to establish himself before daring to ask the paterfamilias to hand over a daughter to his care, something unlikely to happen much before age twenty-five or thirty. Zweig recounted the agonizing secret toll that sexual hypocrisy exacted on his generation, with the constant fear of venereal disease and other sources of shame. "If I try to remember truthfully," he wrote, there was not one of his youthful comrades "who did not at some time look pale and distracted—one because he was sick or feared he would fall sick, another because he was being blackmailed over an abortion, a third because he lacked the money to take a course of treatment without his family's knowledge, a fourth because he didn't know how to pay the alimony for a child claimed by a waitress to be his, a fifth because his wallet had been stolen in a brothel and he dared not go to the police."[39]

The Austrian solution for dealing with all such uncomfortable social realities was, as the Viennese playwright Johann Nestroy observed, to feign comfortable ignorance. "Life was pleasant and amusing," summarizes one historian, "because the Viennese had a superb gift for looking away."[40] Johann Strauss the Younger's 1874 operetta *Die Fledermaus*, his greatest success and the most thoroughly Viennese of all operas, ends Act I with a song that declares, *Glücklich is, wer vergißt, / Was doch nicht zu ändern ist*: "Happy is he who can forget what cannot be changed."

Vienna was accordingly "a city of actors," in a nation suffering from an identity crisis. In Robert Musil's words, "In this country a person always acted differently from how he thought, or thought differently from how he acted. Uninformed observers mistook this for charm." The disconnect between appearance and reality, between surface order and correctness and underlying chaos and deceit, was perfectly reflected in the famous Viennese *Schlamperei* that permeated nearly every aspect of public life. A widely traveled foreign correspondent for *The Times* of London, Wickham Steed, lived in Vienna for a decade at the turn of the century and remarked that no other place he had ever been had so many words to express the idea of "slovenliness." *Schlamperei*, the shrugging off of a regulation or leaving some business half done, was in another sense, however, the only way to get *anything* done in a society groaning under the weight of formal strictures and a bureaucracy whose motto was always, "We can wait." Only the loopholes made life bearable.[41] The socialist leader Viktor Adler summed up Austria's government as *ein durch Schlamperei gemilder-ter Absolutismus*—an absolutism tempered by *Schlamperei*.[42]

The concomitant of *Schlamperei* was the equally venerable Austrian pastime of grumbling, carping, and grouching, which had its own large number of synonyms, especially colorful in the local Viennese dialect. The particular form of barbed invective known as *Wiener Schmäh* was, like *Schlamperei*, a kind of safety valve lurking below the surface charm and *Gemütlichkeit* of Viennese society. "The Viennese does not abuse a person because he dislikes him," explained the Austrian writer and playwright Hermann Bahr, "nor does he abuse him because he wants to scare him off. He abuses him because it's healthy, because it's good for him, and he even likes the person who affords him this plea-sure." ("While they are grumbling they are content," remarked Franz I, whose secret police compiled a dossier for him on every single one of his subjects. "It's only when they fall silent that they are dangerous.")[43]

But the only result of all such evasion and compensation was to constantly sidestep every hard decision. The only Austrian govern-ment policy was usually to have no policy at all beyond "muddling through" (which also had its own Austrian word, *Fortwursteln*), while unresolved and even unrecognized social and political demons festered and grew.[44]

No demon ate at Austrian society more than anti-Semitism. The term had been coined in 1879 by a German anti-Semite, Wilhelm Marr, who had spent a part of his youth working in Vienna. Meant to sound more ideologically respectable than "Jew hatred," it was quickly adopted as the central political program of Austria's new Christian–Social move-ment. Appealing primarily to the Catholic lower middle classes who felt most threatened by the economic and social ascendance of the Jews, the Christian–Social party enjoyed a meteoric rise in the 1890s with their discovery that fanning the flames of anti-Jewish resentment could also be the one great unifier in polyglot Austria.

Karl Lueger, the party's charismatic leader, adeptly rode the tide of anti-Semitism to become the hugely popular mayor of Vienna, serving from 1897 until his death in 1910. Lueger's anti-Semitism was as fine a specimen of Austrian opportunism as the rich history of that phenomenon affords. To the rapt lower-class crowds he joked with easily in Viennese slang—jeering at the "professors" and "scientists" and "lawyers," mocking everything sophisticated or intellectual that made them feel uneasy or intimidated, reassuring cobblers, tailors, greengrocers, and coachmen that they could teach the educated classes a thing or two—he denounced the "money and stock exchange Jews," "press Jews," and the "Judeo-Magyars" from "Judapest," whose push for economic liberalism was ruining "laboring Christian *Volk*" of all classes and

Karl Lueger, Vienna's popular mayor and pioneer anti-Semite

nationalities. He promised to liberate them from "the shameful fetters of Jewish servitude," from those "beasts of prey in human form" who made wolves, lions, and panthers look human by comparison. When confronted with his own continuing and cordial business and social relations with a good number of Vienna's prominent Jewish business leaders, Lueger responded with a famous bit of flippant, Viennese-accented double-think, *Wer a Jud is, bestimm' i'*—"I decide who is a Jew." Lueger's cynicism and hypocrisy in exploiting the ugliest lurking hatreds out of political expediency prompted one of those Jewish associates to rebuke him, "I am not blaming you for being anti-Semitic. I am blaming you for not being anti-Semitic."[45]

Though Hitler dismissed Lueger as "half-hearted" for not embracing the racially based anti-Semitism of another pioneering Austrian populist, Georg von Schönerer, he related in *Mein Kampf* how much he had learned during his years in Vienna from watching the mayor's skill in flattering the urban proletariat, and in understanding that the less propaganda is based on intellectual appeals and reason, and the more on "the emotions of the masses, the more effective it will be."[46]

The rise of anti-Semitism and the failure of liberalism in 1890s Austria were closely entwined. Both Jews and liberalism became victims of their earlier swift successes. The English-style Liberal party that had dominated Austrian politics for the first decades following the 1867 constitution had advanced economic liberalism and expanding democracy—both of which now unleashed forces that turned against them. Only 6 percent of Austrians had the right to vote in the 1860s. But the subsequent expansion of the franchise empowered populist and nationalist movements whose competing demands could not be met. The Reichsrat dissolved into noisy brawls in which competing parties flouted the rules of procedure, staged marathon filibusters, and smuggled in cowbells, harmonicas, whistles, and drums to drown out members of the opposing parties when they tried to speak. "The advent of brutality into politics chalked up its first success," Stefan Zweig wrote of those events of the 1890s. "When the underlying rifts between races and classes, mended so laboriously during the age of conciliation, broke open they widened into ravines. . . . In fact, in that last decade before the new century, war waged by everyone against everyone else in Austria had already begun."[47]

FRÖHLICHE APOKALYPSE

Austria's liberal intelligentsia reacted to the disaster of liberalism's collapse in a characteristically Austrian way, retreating ever further into the life of art, aestheticism, and a kind of high-minded individualism, what the writer Hermann Broch termed the "flight into the apolitical."[48] The incredible blossoming of modernism in 1900 Vienna was the upside of this retreat into the self: the spare architectural functionalism of Adolf Loos, which rejected Vienna's cloying historicism that had elevated appearance over practical use (Loos famously equated "ornament" with "crime"); the atonal inventions of Arnold Schönberg and the spiritual romanticism of Gustav Mahler; and the artistic revolt led by the members of the Vienna Secession, the iconoclastic Gustav Klimt most strikingly, who walked out of the stodgy Academy to create their impressionistic canvases of light and geometry and to pursue the idea of "total art," an entire harmonized other-world of painting, architecture, furniture, and graphics.

Its undercurrent was a plunge into escapism in art and literature that was abandoning society altogether. The mass political movements had undermined "traditional liberal confidence in its own legacy of rationality, moral law, and progress," leaving the Austrian intellectual class "strangely suspended between reality and utopia," said the historian Carl Schorske. Possessed of "both the self-

Portrait of Ludwig Wittgenstein's sister Margaret Stonborough, by Gustav Klimt

delight of the aesthetically cultivated and the self-doubt of the socially functionless" it was reduced to "a culture of sensitive nerves, uneasy hedonism, and often outright anxiety."[49]

A style of intellectual flippancy reinforced the prevailing air of passivity, indifference, and detachment. "The situation is hopeless but not serious," became a catchphrase. The *Kaffeehausliterat* had always been an aesthete, but now he lived in a world almost entirely of his own invention. "A *Kaffeehausliterat*," explained the Austrian Jewish journalist Anton Kuh, himself an incarnation of the type, "is a man who has time to ponder in the coffeehouse what others are not experiencing outside its walls."[50] The elevation of passivity to a national art form was distilled by Robert Musil in *The Man Without Qualities*:

Here one said *es ist passiert*—"it happened"—when people anywhere else would have thought something momentous had taken
place; it was a curious expression, existing nowhere else in German
or in any other language, a whiff of which could make facts and
catastrophes as light as eiderdown and thoughts. So yes, perhaps in
spite of so much that argues against it, Kakania was after all a country for geniuses; and it is probably for that reason it went under.[51]

A great influence on the thinking of the intelligentsia of "Young
Vienna" was the philosophical ideas of the physicist Ernst Mach,
named in 1895 to the newly created Chair for History of the Induc-
tive Sciences at the University of Vienna. Rejecting all a priori truth,
Mach took empiricism to new extremes: knowledge is based only on
appearance, he insisted, and appearances are nothing but "the world
of our own sensations." Thus, there are no such things as objects, per-
manence, or even reality in any objective sense. Mach's public lectures
were avidly attended by the denizens of Young Vienna, who credited
his ideas as a wellspring of much of their artistic sensibility. The poly-
math Egon Friedell—music critic, philosopher, cabaret performer,
playwright, among other things—asserted that in choosing to depict
light over *things*, Klimt and his fellow modernists had simply "painted
Mach." Mach's idea that reality was nothing but illusion was in perfect
keeping with the *Zeitgeist*.[52]

And meanwhile, Austria hovered on the edge of disaster. At the
end of *The Emperor's Tomb* Roth's Polish count, as he contemplates
the impending dissolution of the empire, bangs his fist furiously on
the table and exclaims, "You people are responsible for all this—you
brought down the State with your idiot coffeehouse witticisms." Hermann Broch termed that contradictory period of creative brilliance
and intellectual frivolity of the empire's last decade, which others more
poetically called "Vienna's Golden Autumn," the *fröhliche Apoka-
lypse*—the "Joyful Apocalypse."[53]

In July 1914, in the days just before the outbreak of the First World
War, Karl Kraus more bitterly described his country as "the Austrian
experimental station for the end of the world." He was referring to the

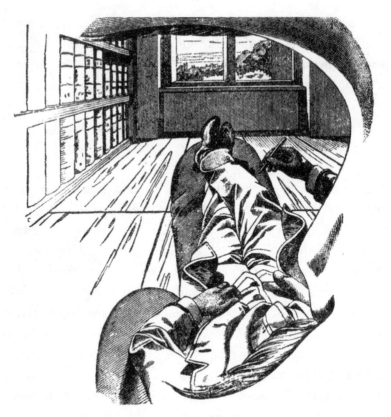

Ernst Mach's ironic self-portrait, *Self-Inspection of the Ego*

explosive force of nationalism within its borders that was now about to be replicated on a global scale. Less noticed at the time was the particularly lethal concoction that Austrian society proved could be brewed up from a mixture of charm, frivolity, passivity, and opportunism, with a dash of casual cruelty. "Like you I feel an unrestrained affection for Vienna and Austria," Sigmund Freud told a young visitor who called on him at his office in November 1918. "Although perhaps unlike you, I know her abysses."[54]

· 2 ·

Alle echten Wiener sind aus Brünn

THE MANCHESTER OF MORAVIA

The direct source of the secure prosperity that enfolded Gödel's youth both before and after the First World War was the flourishing textile industry of Brünn, the city where he spent the first eighteen years of his life. Capital of the province of Moravia and the second largest city in the lands that would form the new Czechoslovak Republic, Brünn prior to the industrial revolution had been a small medieval fortress town known mainly for its heroic defense against besieging Swedish troops in 1645, and for its proximity to the more typically Austrian ignominious military defeat at the Battle of Austerlitz in 1805, which ended the Austrian emperor's claim to the crown of the Holy Roman Empire. But in the ensuing decades, Brünn became one of the first beneficiaries of the Austro-Hungarian Empire's slow industrial awakening, aided by its location just seventy miles north of the imperial capital at the confluence of two tributaries of the Danube that fed the millraces of its first factories.

By the mid-nineteenth century Brünn had established itself as a major center of fine woolen production, with two dozen large mills, now powered by smoke-billowing steam engines, that already were rivalling England's in the quality and price of their products, securing for the city the name "the Moravian Manchester." In the first decade of the new century Brünn's woolen mills employed more than 13,000 workers, with thousands more in the associated garment factories and in the heavy machinery works that churned out steam engines, rail cars, boilers, turbines, and electrical equipment to meet the burgeoning demands of the local industrial revolution.[1]

The clustered smokestacks of its factories filled the grittier industrial suburbs and the part of the city known as Altbrünn, a small crescent of low land lying between the Schwarzach River and the abrupt heights of the formerly walled central town. On the hillside above, floating ethereally over the noise and smoke of nineteenth-century progress, was the fourteenth-century Cathedral of Saints Peter and Paul, with its recently added 275-foot-high Gothic-revival twin spires; and towering above it all, on its own solitary promontory, stood the forbidding ramparts of Spielberg Castle, once the most notorious citadel of the Habsburg Empire, with its dungeon cells where political prisoners were fastened to dank walls with iron chains.

A gritty view of industrial Brünn, 1915

Brünn

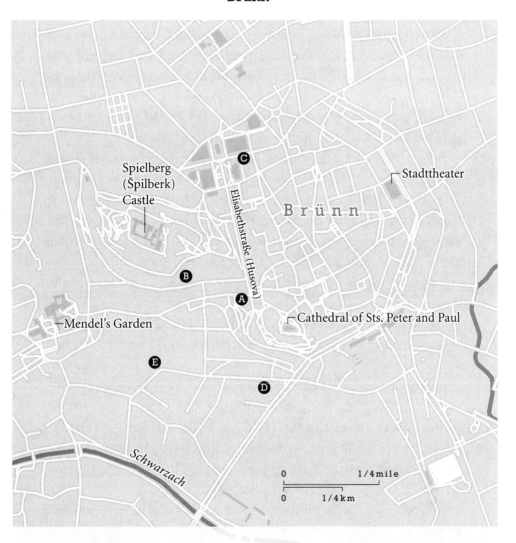

Ⓐ birthplace, Bäckergasse (Pekařská) 5
Ⓑ family villa, Spielberggasse (Pellicova) 8a
Ⓒ Evangelische Privat-Volks- und Bürgerschule
Ⓓ Staats-Realgymnasium
Ⓔ Friedrich Redlich Feintuch- und Schafwollwaren Fabrik

Stretching to the south was the valley formed by the Schwarzach River as it flowed to join the Thaya and then the March on their way to the Danube, carving out an ancient natural pass that connected to Vienna through the Carpathian Mountains—a bright green ribbon of orchards and vineyards, traversed since 1839 by the Emperor Ferdinand Northern Railway, the first in the Austrian Empire.

By the year of Gödel's birth in 1906, the population of Brünn was approaching 120,000, nearly two-thirds of whom were ethnic Germans. From their comfortable villas and apartment houses along Brünn's own miniature ring street, built like Vienna's upon the angled segments where medieval walls had stood, the prosperous German middle class dominated the town hall and the economic life of the city, looking down—literally and figuratively—on the Czechs who filled the ranks of domestic servants and factory employees. Unlike the predominantly German regions of Bohemia and Moravia immediately adjacent to Germany and Austria that German nationalists had already dubbed the "Sudetenland" in their hopes of annexing them to an eventual, greater German state, Brünn was a German-speaking "language island" surrounded by almost entirely Czech suburbs and towns. During the Second World War, the Nazis would try to imbue that shaky status with heroic overtones, trumpeting Brünn like many other similarly situated cities as *ein deutsches Bollwerk*, "a German bastion."[2]

In Gödel's youth the term could have been apt as well. "The intelligentsia and the 'upper ten thousand' " were all German, recalled his older brother, Rudi, of the Brünn of their boyhood days; so were the dominant cultural life and the most prestigious schools.[3] Gödel's mother was wholly typical of that class in the casual dislike she expressed for the "Slavs." If forced to communicate in Czech, most Germans of Bohemia and Moravia resorted disdainfully to a kind of pidgin Czech, "in which badly pronounced words were fitted into German syntax."[4] Street brawls between young toughs occasionally required the dragoons to be called out to separate German and Czech rioters—Gödel's mother remembered the clatter of hoofs on the cobblestones at night in front of the house—but there was little doubt who stood where in the social pecking order, at least before the founding of the Czechoslovak state in 1919.

The apartment house where Gödel was born at Bäckergasse 5 on April 28, 1906, stood just off Elisabethstraße, the segment of the ring street that ran past the base of Spielberg in the prosperous German part of town. Both of his parents had grown up next door at Bäckergasse 9, built in the "Biedermeier" style typical of apartment buildings of mid-nineteenth-century bourgeois Vienna, with a large open central courtyard—where, his mother recalled, the neighbors would gather together at the end of the day, often at the sounding of Taps from the barracks of the Spielberg above. The

"Herr Warum" as a baby, with brother Rudi

families often joined together as well to celebrate holidays, to perform amateur theatricals, or for musical evenings; Leo Slezak, a contemporary of Gödel's parents who would become a famous operatic tenor, was a neighbor in the building.

Like his future father-in-law, who had slowly worked his way from bitter poverty to the top ranks of management in the local branch of the House of Schoeller, whose textile manufacturing empire spanned Austria and Germany, Rudolf Gödel had determined at an early age to make his fortune in the town's trademark industry. In 1901 he was a rising manager at the Friedrich Redlich Fine Fabric and Woolen Goods Factory, Brünn, when at age twenty-seven he married Marianne Handschuh. She was twenty-one.

There were shades of Kurt in both his parents. His brother Rudi would recall their father as "rather ponderous and serious minded," "an energetic, efficient, and thoroughly practically minded man" who made up for a lack of personal warmth with sympathetic indulgence of his wife and children. The marriage was no love match, Rudi later saw, but "was certainly based on sympathy and affection." As a father, Rudolf Gödel likewise "fulfilled many wishes for us children and later generously supported our studies in Vienna," despite his own poor showing as a scholar in his youth. Sent to the Gymnasium in Brünn, he had evinced "no talent and no interest" whatever in the classical education in Latin and Greek it offered, and switched at age twelve to the town's industrial training school, from which he was hired immediately upon graduation by the Redlich firm.

Marianne, despite her apparent contentment in running a household and seeing to the needs of the family, was much more the intellectual of the two, as well as being a mother who believed in the simple and congenial theory of being a friend to her children—a role made easier by being able to turn over the actual work of cooking, cleaning, and childrearing to nurses, servants, and governesses. Rudi recalled the

The Gödel family: Marianne, Kurt, Rudolf, Rudi

"warm, almost comradely" relationship both sons enjoyed with their mother that would continue throughout their adult lives.

Marianne's father had always retained a hard and rigid side from the rigors and poverty of his youth: his father had been a hand weaver put out of work by industrialization, and in his grim determination to succeed the young Gustav Handschuh had had to acquire his education alongside his work, often copying out by hand the texts of borrowed books he could not afford to buy. But Marianne, perhaps in reaction, had grown up a lively and cheerful girl, good at gymnastics and ice skating, an excellent pianist who enjoyed accompanying friends and family members in Schubert and Strauss lieder, a voracious reader of the great works of German literature who throughout her life could recite by heart the poems of Goethe and Heine—Rudi remembered his mother's treasured library of seventy volumes of classic German literature, many of them first editions—and with an intellectual confidence strong enough to form and defend her own conclusions, uninfluenced by general opinion.

The middle-class insecurities that occasionally threatened to intrude did so only in discreet silence. Rudolf Gödel's father had committed suicide at a young age, a fact simply never mentioned in the family. Rudolf had then been sent back from Vienna to Brünn to live with his father's sister Anna. He later broke with his mother entirely, when she and her relatives began importuning their increasingly successful young relation for monetary assistance. There were similar strains on Marianne's side of the family, including a wastrel brother who had had to leave the army after running up debts and squandering a considerable sum loaned him by Rudolf, which he never repaid, and another brother who resented Rudolf's successes—"which was why my father cultivated contact with almost no relations," Rudi later understood.

In 1913, Rudolf Gödel purchased a plot of land on Spielberggasse backing up to the wooded slopes of the castle summit, their own small bastion of bourgeois family security against a changing world. The villa he built there on the eve of the First World War was three stories tall, with a cheery succession of rambling gables, filled with light from windows that overlooked a large garden in the rear, a third of an acre

The Gödel villa on Spielberggasse (now Pellicova)

filled with fruit trees that bloomed magnificently in the spring, and to the front commanded a view of the city and valley below. The centerpiece of the living room was a grand piano made by Vienna's finest maker, Bösendorfer, the official piano maker to the emperor. A large entrance hall on the first floor was paneled in natural wood and furnished in modern good taste with Jugendstil furniture, the seats of the chairs covered in the Arts-and-Crafts prints of the turn-of-the-century Viennese movement that evoked and celebrated the handcrafts of the pre-industrial age that Brünn had led the world in destroying.

HERR WARUM

Science and progress were in the very air of Brünn. A ten-minute walk down the hill from the Gödel villa was the Augustinian abbey of Alt-brünn, where Gregor Mendel spent a decade hybridizing and meticulously sorting and counting the blossoms and pods of 10,000 pea plants, laying the foundation of modern genetics. The 1910 dedication of a marble statue of Mendel, erected "by the friends of science" in the square in front of his monastery garden, brought famous biologists from across Europe to the city. (In 1950, during the heyday of Stalin's promotion of the anti-genetics theories of the Soviet agronomist Trofim Lysenko, the Czech Army hauled the statue off in the middle of the night and deposited it unceremoniously at the back of an inner courtyard of the monastery. Only in 1965, after the fall of Nikita Khrushchev, who had also lent support to Lysenkoism, was the statue restored to the dignity of a more prominent spot in the monastery garden, where it stands today.)

Ten minutes in the other direction was the Brünn City Theater, the first in Europe to be equipped with electric lights when it opened in 1883, the forward-looking town council having negotiated the contract for their design with Thomas Edison, who visited the city in 1911 to view his creation. Brünn's streetcars were electrified in 1900, a year before Manhattan's.

Kurt Gödel's family would always claim to have anticipated his future genius in the nickname they bestowed on him when he was four years old: *Herr Warum*, "Mr. Why." As Rudi recalled, "he always wanted to get to the bottom of everything through particularly intensive questions." Gödel himself added a slightly sharper and more defiant edge to a cute family story: As he told his psychiatrist in Princeton a half century later, he had been as a child "always curious, questioning authority, requiring reasons," and these traits would become a source of adolescent friction with his father when, often, he chose to remain at home immersed in a book rather than joining the rest of the family on their ritual Sunday outings.[5]

"The highest aim of my life (conceived in puberty)," Gödel wrote many years later, "is pleasure of cognition." He recalled the moment that first kindled his passionate interest in science, and the vivid picture forever associated in his mind with the place where, at age fifteen while staying with his family at a resort town in the mountains of western Czechoslovakia, he had picked up a biography of Goethe. He wrote his mother in the summer of 1946,

The book "Goethe" by Chamberlain that you write about carries with it many youthful memories for me. I read it (oddly enough exactly 25 years ago) in Marienbad and see before me still today the strangely purple-colored flowers that everything was covered in. It is unbelievable how something like that leaves an impression on you. . . . This Goethe book was the beginning of my interest in Goethe's science of colors and his argument with Newton, and thus indirectly led to my choice of profession. Thus curious threads spin themselves through life that one doesn't discover until one grows older.

Vacationing in 1941 in Brooklin, Maine, he had seen the very same flowers: "How strangely it moved me," he wrote.[6]

The nostalgic haze that enveloped his childhood would remain with him through his life. If it was intensified by the years of personal and political turmoil and upheaval that would follow, it was nonetheless a genuine reflection of a happy and safe world. The two brothers always got on well, their four-year age difference helping to ease the natural rivalry, and they spent hours in the garden playing with the two family dogs, a Doberman and a terrier, planting a small vegetable plot under their governess's supervision, exploring the castle grounds and playing on its snowy hillside in winter ("Yes, sledding was my passion back then, I still recall it well," he wrote his mother fifty years later), and sharing make-believe games of their future careers, which Gödel still fondly recalled to his brother as he approached his fiftieth year:

Do you remember our city games in which you were a factory owner and I was the mayor? Back then it seemed so impossible to me that we would really see the dates of years such as 1950. In the city games we were always both already rich by this time. That has not happened yet, but we also cannot complain that we are bad off.[7]

The attic was filled with scientific and expensive toys—a telescope through which they could see the detailed stone carvings on the Gothic dome of the cathedral, toy trains, an army of eight hundred tin soldiers with six matching wagons and horses. They were always allowed to pick a gift at Christmastime from the toy catalogues of one of Vienna's fancy toy stores, Mühlhauser or Niessner, a procedure their mother thought "unpoetic," but which also would be etched in Gödel's lasting memories of childhood pleasures. Decades later, when his mother once inquired what he might like for a present, he asked if she could just send him a toy catalogue from one of the same shops, or their successors: "It would interest me greatly to see what kind of progress the toy industry has made in the last 45 years," he told her, adding in a wry allusion to his own collection of toy armaments in a far more innocent time, "Are there not also already little atom bombs for children?"[8]

In that world of scientific wonders and material progress, barbarity and superstition seemed to be swiftly becoming relics of the past. The only religious instruction the Gödel children received was in classes at the elementary school Kurt attended starting at age six, the Evange-lische Volksschule. Marianne Gödel had been brought up in a Protes-tant family where, in Rudi's words, "enlightened piety" reigned; their father was nominally a member of the Old Catholic order (a group of mostly German and Austrian believers who had broken with the Roman Catholic Church in the late nineteenth century over the doc-trine of papal infallibility), but except for the occasion of his marriage he had scarcely set foot in a church since reaching adulthood. It is vaguely possible that he was of distant Jewish ancestry: a number of Gödels would appear on the lists of Jews rounded up by the Gestapo in Moravia, and likewise a great many Goedels in Holland.

Both Kurt and his brother would later express a certain regret about their cursory exposure to religion in that freethinking household. "How necessary the support is that is thereby taken away from one is some-thing I first felt later when deaths and blows of fate hit our family," Rudi ruefully remarked in his eightieth year.[9] One of the few aspects of human nature that could reliably arouse Kurt Gödel's undisguised contempt throughout his life were the doctrinal absurdities and inven-tions of formal religion, which he saw as a hindrance to true religious feelings. "Is it true that the Resurrection was celebrated in Brünn on

Holy Saturday??" he once wrote his mother in amused bewilderment. "In a Lutheran church in Brünn anything may be possible, but *that* would really surprise me."[10] And when she asked her son whether it was true, as she had read, that it was the religion classes at the Catholic elementary school Einstein had attended that had inspired some of his later scientific ideas, Gödel replied:

> What you write about Einstein's biography is correct, so far as I know, . . . that through Religion class the foundation was laid for his seeking a unified theory for the entire universe. That really must have been a very good and interesting Religion class. Because with the kind we had, that would certainly not have been possible.[11]

"Religions are for the most part bad," Gödel wrote in a notebook many years later, "but not religion."[12]

He underwent his own nominal ceremony of baptism in the Lutheran church in Altbrünn. Acting as his godfather was his father's employer and later business partner, Friedrich Redlich, the second member of his family to bear that name and ownership of the family woolen mill. Like many of the city's leading industrialists, he was of Jewish ancestry, part of the rising tide of Jewish success following the abolition in 1867 of the last legal restrictions on Jews' professions and places of residency in the Austrian Empire. Joseph II had taken the first steps toward encouraging Jews to establish factories in Moravia, along with advancing their greater integration into society by requiring Jews to send their children to schools using German as the language of instruction. But he had retained some of the onerous restrictions of his predecessors, particularly the notorious *Familiantengesetze*, which permitted Jews in Bohemia, Moravia, and Silesia to marry only if they possessed a state-issued number, which could be passed on to the eldest son only after the death of each holder. For four centuries Jews had been legally barred from living in Brünn altogether; that restriction, too, was abolished only in 1848.[13]

But by 1900 a flourishing Jewish community of about eight thousand, making up some 10 percent of the German-speaking population, lived in Brünn. Nearly all strongly identified with German language and culture, and many like Friedrich Redlich had come to regard conversion to Protestantism as a final step in assimilation that they believed would permanently secure their place in society. His son Fritz Redlich, thirteen years older than Kurt, would learn just how ephemeral that place was when the German Army marched into Czechoslovakia in March 1939 and declared the Reich's "Protectorate" over the rump states of Bohemia and Moravia. Neither his military service in the Austrian Army during the Great War, for which he was awarded the Iron Cross for gallantry, nor his Protestant faith would save him from the Nazis, whose snide term *stehend getauft*—"standing baptism"— gave the back of the hand to Jews like his father who had converted as adults. Like the ten thousand other Brünnites whom the Nazis racially categorized as Jews, Fritz Redlich would be transported to the concentration camp at Terezín in 1942, and from there to his extermination at Auschwitz two years later. By then Gödel's nostalgic memories would be the only remnant of his bright and peaceful boyhood days.[14]

A "SORT OF" SWITZERLAND

The lines of the new Czechoslovak state drawn by the victorious powers at the 1919 Paris Peace Conference left 3 million Germans on the wrong side of the border. President Woodrow Wilson's declared principle of national self-determination had left open the possibility of small adjustments in the interests of national and economic viability, and the Czech delegation had cannily made the most of it, lobbying heavily to have all of the lands within the traditional and historic borders of Bohemia and Moravia included as part of the new Czechoslovak Republic. To blunt the appeal that Germans living in these regions had sent directly to President Wilson, invoking their own right of self-determination to opt out of the new state, the Czech delegates spoke reassuringly of making Czechoslovakia into a "sort of Switzerland" (or a "Switzerland of the East"), in which all nationalities would enjoy equal rights within a national federation.

Back home, the message was starkly different. Alois Rašín, one of the prominent leaders of the provisional government, refused to discuss German demands for autonomy, disdainfully declaring, "I don't negotiate with rebels." Tomáš Masaryk, the revered founding father of the Czechoslovak state, pointedly branded the Germans of Bohemia and Moravia as "settlers and colonists." Urged on by the sentiments of their leaders, crowds in cities across Bohemia, Moravia, and Slovakia tore down the emblem of the imperial double eagle from public buildings and schools and toppled and carted off statues of Habsburg rulers and military heroes. Municipalities quickly set to work renaming streets and squares with Czech names. In Brünn—now Brno, under its new Czech name—the local government annexed twenty-three surrounding municipalities, nearly doubling the population of the city in order to lock in place a permanent Czech majority.[15]

The lines of ethnolinguistic identity in the region had ironically been sharpened by the old empire's last great attempt to defuse nationalistic tensions, the Moravian Compromise of 1905, which allocated seats for separate Czech and German voting blocs in the provincial legislature. But that, as the historian Pieter M. Judson notes, had had the perverse consequence of "forcing people previously indifferent to nationalist identifications" to choose one or the other: citizens who tried to register under the new law as "Austrian," "Catholic," or "Habsburg loyalist" were forced to identify themselves now only as "German."[16]

The powerfully reinforced sense of German ethnic identity, exacerbated by provincial insecurity, had accordingly made the Sudeten Germans the most German-nationalist group of all the Habsburg Empire. They had volunteered and fought in the Great War with a devotion that surpassed even the German Reich itself, suffering losses of more than 100,000—a rate of 34.5 killed per 1,000 population, 25 percent greater than that of Germany and Austria as a whole.[17]

While granting a complex set of minority language rights, the new Czechoslovak government built on the 1905 precedent by giving its census takers the power to "correct" a person's declaration of nationality based on "objective" characteristics, and imposing a heavy fine, equal to as much as a week's wages for the average worker, on any person who "consciously" misrepresented his nationality. Some 400,000 ethnic

Germans were changed to Czechs in the 1921 census by these methods, resulting in the appropriation of German schools and other erosions of the German community's position. The early years of Czechoslovak nation-building were inextricably bound with the throwing-off of any vestiges of German oppression: it was not until 1926 that a representative of one of the German parties in parliament was given a ministerial portfolio in a coalition government.[18]

All of this sat especially hard with the once-comfortable German elite of Brno. But there were undeniable considerations in favor of trying to make the best of things in their existing home, rather than exercising their rights under the Versailles Treaty's Option Clause, which allowed citizens of any of the regions of the former Austro-Hungarian Empire to claim citizenship in any of the successor states that better accorded with their "race" or "nation." (That was an option infamously denied to 75,000 German-speaking Jews living outside the new borders of the Austrian state, whom the government ruled did not qualify for Austrian citizenship because they were not "racially" German.)[19]

In contrast to Austria's plunge into economic collapse, Czechoslovakia had survived the ravages of the war relatively unscathed, the richest of the successor states to emerge from the shattered fragments of the Habsburg Empire. Although wartime food rationing and shortages continued for several years after the war in Czechoslovakia as they did in Austria, the situation was far less dire there. Along with abundant supplies of coal and large tracts of rich agricultural lands, making it the most self-sufficient in food of the successor states, Czechoslovakia's territory encompassed two-thirds of the entire industrial capacity of the former Austrian Empire and its steel industry outproduced all of the other states of Central Europe combined.[20]

The Austrian delegation that had hopefully arrived in Paris, representing the provisional government of the new Austrian Republic proclaimed the day after the Armistice, had vainly tried to make the case that, as a new country like the other successor states of the Habsburg monarchy, Austria alone should not be saddled with reparations that were properly borne by all the regions that had fought on the side of Austria-Hungary. Instead, the delegation found itself confined like prisoners in a small palace surrounded by barbed wire, and ignored. As in Germany, hyperinflation was ravaging the Austrian economy, reducing the value of the Austrian krone to 1/13,000th its prewar level, while an embargo on coal by Czechoslovakia left Austria desperately short of fuel for heat and transport: along with much of its food, Austria had imported 99 percent of its coal from other parts of the empire.[21]

In 1919, Rudi Gödel moved to Vienna to begin his medical studies at the University of Vienna. Famine and tuberculosis gripped the city. Stefan Zweig described in his memoirs what it was like to travel to Vienna, the capital of "the grey, lifeless shadow of the old Austro-Hungarian Monarchy," in those first years of gripping shortages following the war's end:

> A journey to Austria at that time called for the kind of preparations you would make for an expedition to the Arctic. You had to equip yourself with warm clothing and woolen underwear, because everyone knew there was no coal on the Austrian side of the border—and winter was coming. You had your shoes soled; once across the border the only footwear available was wooden clogs. You took as much food and chocolate with you as [they] would allow you to take out of the country, to keep from starving until you were issued with your first ration cards for bread and fats. Baggage had to be insured for as high a sum as possible, because most of the baggage vans were looted, and every shoe or item of clothing was irreplaceable.[22]

The train trip between Vienna and Brno, which normally took two or three hours, sometimes took twelve or even fifteen. Rudi's parents came whenever they could, bringing food and household supplies from far richer Czechoslovakia to keep him fed and clothed.

Although their father had lost most of his savings in war bonds whose value was wiped out, within a year the quickly recovering economy of Czechoslovakia had restored much of the family's fortunes. Gödel would later describe his family as "almost wealthy": in 1920 his father bought one of the first automobiles in the country, a light blue-grey Chrysler complete with chauffeur to drive the family on its weekly Sunday outings to the countryside. As before the war, they took summer holidays at the posh resorts of the old empire in Marienbad (now Mariánské Lázně) and Franzensbad (Františkovy Lázně) in the mountains of western Bohemia, and twice even to Abbazia (Opatija) on the Adriatic. Their mother, Rudi recalled, "led the life of a society lady," going to the theater, meeting her girlfriends at cafés, dining at the Grand Hotel opposite the railway station, attending musical recitals and song evenings.

But there was no question that Kurt's and Rudi's futures lay in Vienna, even if their father's fortunes remained tied to his once-more thriving factory. They were part of the inevitable exodus of young Czech Germans who doubted their future in a nation that had chosen to define its existence not merely in terms of independence from the Austro-Hungarian Empire, but as an ethnolinguistic triumph over their one-time oppressors. "I am a true Viennese," their fellow Brünnite Leo Slezak would quip in his memoirs. "All true Viennese are from Brünn."[23]

KURT IS "GOOD" AT MATH

In his four years at the Evangelische Volksschule, Kurt had frequently been absent due to illness. At age eight he suffered a severe case of rheumatic fever, which convinced him for the rest of his life that he had a permanently weakened heart. The incident, in his brother's judgment, may have been the start of his lifelong hypochondria. Several years later he underwent an appendectomy, which he recovered from without complications.[24]

Even so, he maintained almost perfect grades—the proverbial *lauter Einser Kind*, the "nothing but ones child," receiving the highest mark of 1 on a 1-to-5 scale on his report cards. His Volksschule completion certificate in July 1916 awarded him the highest commendation ("*sehr gut*") in all areas—religious education, language, arithmetic—while noting his "commendable" behavior.[25] That fall he began his studies at the K. K. Staats-Realgymnasium mit deutscher Unterrichtssprache in

The Realgymnasium in Brünn, left; down the street is the
smokestack of the Redlich factory

Brünn, the "Imperial-Royal State Realgymnasium with Instruction in
the German Language."

The Realgymnasium was a product of the nineteenth-century education reforms of the Austrian Empire, adding a third alternative for
high-school-level education. Previously, the only choices were the intensive classical curriculum of the Gymnasiums, which had originally
provided the only route to admission to a university, and the more voca-
tionally oriented *Realschulen*, which were intended to provide basic
competency in writing, foreign languages, and mathematics for clerks,
technical workers, and others with no intention of pursuing a scholarly
career or learned profession. Like the Gymnasiums, the Realgymna-
siums were a university preparatory course, but with an emphasis on
science and modern foreign languages over Latin and Greek.

The Realgymnasium in Brünn stood on Wawrastraße (now
Hybešova), a ten-minute walk from Spielberggasse descending the steep
hill into the industrial part of town. The Redlich factory was just a

few blocks down the same street, its tall smokestacks clearly visible in a photograph taken at the time from the street corner in front of the school.

Gödel's eight years at the Realgymnasium included classes in Latin, French, German, mathematics, physics, chemistry, geography and history, natural history, freehand drawing, and religion, as well as electives in English and in Gabelsberger shorthand.[26] Both Kurt and Rudi also studied English with a private tutor hired by their father.

Kurt did not bother taking an offered class in Czech language, choosing the shorthand class in his senior years instead. A school classmate and friend, Harry Klepetar, recalled that Kurt was the only one of the students at the school who did not seem to know even a few words of the local language. He did not however share his mother's ill-will and prejudice toward the Czechs. Many years later he reproved her for her attitude, saying, "You are astonished that I find the Slavs likeable, but you yourself give an example in your last letter of how their supposedly unlikeable qualities are based on slander"—she had accused some unidentified Czechs of cutting down some trees in the garden of the villa during the Second World War, though later admitted that their own housekeeper was to blame. But his choice of electives was a small yet definite statement that he was turning his back for good on a future in his place of birth. "He considered himself always Austrian," Klepetar said, "and an exile in Czechoslovakia."[27]

His facility in Gabelsberger shorthand would serve him well in any case. The system was widely taught in Austria at the time, used not only by stenographers and clerks but by many scholars and professionals as a fast and semi-private method of taking notes. The system would be rendered obsolete the year of Kurt's graduation, 1924, when it was replaced by a new "unified" shorthand that superseded the several different systems then in use in Germany and Austria. But, like many of his contemporaries, Gödel would use the system throughout his life, filling dozens of notebooks in Gabelsberger script to record his philosophical reflections, conversations with friends, drafts of mathematical proofs, and some of his most intimate thoughts—anxieties over his social self-consciousness and torment over his perceived shortcomings, sexual doubts, and worries about ever finding a secure position to pursue his scholarly ambitions.

Reflecting the same devotion to education that prevailed throughout the empire, Jewish students (tactfully described as of the *israelitisch* faith in the school's annual report) made up 40 percent of the Realgymnasium's enrollment, four times their proportion among the German-speaking population of Brno at the time.

Gödel's close Jewish friends at school would all remarkably—and exceptionally—survive the war. Harry Klepetar, who became the political editor of the leading German-language newspaper in Prague, the liberal *Prager Tagblatt* which dared to denounce the Nazis, fled for his life to Shanghai, where he lived for nine years before moving to New York in the 1950s and reestablishing contact with his old school friend; he would later work for an organization seeking reparations for looted Jewish property.[28]

Adolf Hochwald, who played chess with Kurt at school, was the only member of his large family to survive the Nazis, managing a perilous escape in 1939 that took him to Switzerland, back to Prague, again to Switzerland, then to Spain, Portugal, Haiti, and Canada, finally making his way in 1941 to Boston, and then to Lake Placid, New York, where he worked as a researcher at the state tuberculosis hospital, before dropping dead of a heart attack at age fifty. He, too, had worked to help the victims of Nazism after the war, as a doctor in the United Nations Relief and Rehabilitation Administration.[29]

And Fritz Löw Beer, a member of a wealthy Jewish family of Brünn industrialists that had made its fortune in textiles and sugar—and whose cousin Greta Tugendhat commissioned the architect Mies van der Rohe to build a landmark modernist house that still stands outside of Brno—was able to make it to New York, where he became a dealer in Asian art and, unusually for that first generation of collectors, a world-renowned authority who published well-regarded scholarly articles on Chinese lacquerware.

Virtually none of Gödel's other Jewish classmates would be so fortunate. "There was a 25th class reunion planned but it looks like it will

K. K. Staats-Realgymnasium mit deutscher Unterrichtssprache in Brünn.

Katalog-Nr. *21.*

Schuljahr 1916/17.

Semestral-Ausweis

für

Gödel Kurt, Schüler der *ersten A. Klasse*

Betragen	*sehr gut.*
Religionslehre	*sehr gut.*
Deutsche Sprache (als Unterrichtssprache).	*sehr gut.*
Lateinische Sprache	*sehr gut.*
Französische Sprache	*♂*
Geschichte	*♂*
Geographie	*sehr gut.*
Mathematik	*gut.*
Naturgeschichte	*sehr gut.*
Chemie	*♂*
Physik	*♂*
Freihandzeichnen	*sehr gut.*
Schreiben	*sehr gut.*
Turnen	*sehr gut.*
Böhmische Sprache (rel. obligat)	
Stenographie (.... Kurs)	
Gesang (.... Kurs)	

Freie Lehrgegenstände.

Brünn, am *10. Februar* 1917.

Direktor.

Klassenvorstand.

Notenskala.

Betragen	sehr gut	gut	entsprechend	nicht entsprechend
Fortgang	sehr gut	gut	genügend	nicht genügend

Gödel's report card, age eleven

come to nothing," Gödel would sadly tell his mother in 1949. "How many are even still alive? Hochwald, with whom I had contact for a while, was very pessimistic in this regard."[30]

The relentless perfectionism in Gödel's mental and emotional makeup was plainly evident in his schoolwork. In every class he received a *sehr gut* on his report cards—with the exception of one *gut* in mathematics, during his first year. Rudi recalled that it had become the lore of the school that his brother was the only student never to have made a single grammatical error in his Latin exercises during his eight years of study.

Harry Klepetar told Gödel's biographer John Dawson that he had considered Brünn's Realgymnasium "one of the best schools in the Austrian monarchy and later in Czechoslovakia." Gödel himself thought differently. When his mother many years later sent him a book about Brünn, he observed drily that it had much to say about the town's Gymnasium, but "regarding the Realgymnasium, *not a* word! Probably its past is insufficiently glorious or even inglorious, which, given the conditions at the time I attended it, would not surprise me at all."[31]

In any case, by the time Gödel was fourteen he was already immersing himself in mathematics and philosophy on his own. He mastered calculus and other university-level topics in mathematics, and read Kant for the first time at age sixteen.[32] His intellectual precociousness, his brother thought, also "perhaps explained why he already felt attracted by older women. An escapade of my brother's as a high school student with a ten-year-older lady caused a commotion in our family." It would not be the last time he caused a commotion on that score.

In 1924 he joined the migration of Czech Germans to the capital of the now forever vanished Austrian Empire, moving into the apartment his brother had rented five blocks from the university and a world away from the provincial town of his youth.

· 3 ·

Vienna 1924

THE HEAD WITHOUT A BODY

"I shall live on with the torso and imagine that it is the whole," Freud said of life in Austria after the war. Many Viennese reversed the metaphor: their city, now an imperial capital minus an empire, was "a head without a body"—and, in the disparaging view of the rest of Austria, a hydrocephalic one at that, a *Wasserkopf*.[1]

Either way its hopes for survival seemed dubious. The coming of peace had done nothing to lighten Karl Kraus's apocalyptic mood: A few years after the war's end he published a bitterly satirical play, *The Last Days of Mankind*. ("The performance of this drama, which would stretch out over some ten days measured in earthly time, was conceived for a theater on Mars," he explained in the preface.) The global influenza pandemic that accompanied the end of the war took the lives of the brilliant artists of the Vienna Secession Gustav Klimt and Egon Schiele, Freud's beloved daughter Sophie, and 24,000 other Austrians along with more than 2 million across Europe.[2]

"The Vienna in which I grew up was a city still pretending to be a world metropolis of cosmopolitan elegance," recalled the writer George Clare, "but behind the imposing façade of former imperial splendor lurked defeat, poverty, and fear. Behind the baroque masonry lay dark, dank corridors filled with the stale smell of over-boiled cabbage and human sweat and the indefinable but clearly discernible odor of hatred and envy." An air of political uncertainty hung in the streets, fueled by a growing conviction across the political spectrum that their new remnant of a state was simply unviable, its only salvation a confederation with Germany sooner or later. The hyperinflation that was turning life savings into pocket change, as basketfuls of notes overprinted with extra zeroes circulated, added to the sense that the props holding up society had been pulled away. "One cannot talk about prices," the German physicist Max von Laue reported from Vienna on a visit in 1922. "Before one's sentence has ended, they have increased again."[3]

One of the few advantages of the inflation of the Austrian krone was that following its replacement in 1924 by the new schilling, at a rate of 1 to 10,000, the socialist government of the city of Vienna imposed rent controls on apartments that set the amount of the new rents in schillings at the same amount as the old rents in krone. The result was "a certain luxury in living space," recalled Karl Menger, son of the economist Carl Menger, and soon to be one of Gödel's closest mentors and colleagues. The apartment that Gödel shared with Rudi at Florianigasse 42 was the first of a series of spacious apartments the brothers would rent, moving each year to yet another solidly built four-story, bourgeois, turn-of-the-century building in the neighborhood of the university. The other consequence of the city's rent control, however, was to exacerbate Vienna's longstanding housing shortage. There was little incentive for owners to construct new buildings, or for that matter to maintain existing ones. Menger one day narrowly escaped death when a half-ton piece of stucco crashed down from the upper story of a house onto the pavement two steps in front of him.[4]

But by the mid-1920s, Menger recalled, "at long last the traditional optimism of the Viennese reappeared." A large loan from the Allies stabilized the currency, and with it came a proviso guaranteeing Austria's independence and commitment not to join Germany. Although Austria was governed from 1920 on by a series of conservative coalitions headed by the Christian–Socials, who continued to

Gödel's Vienna

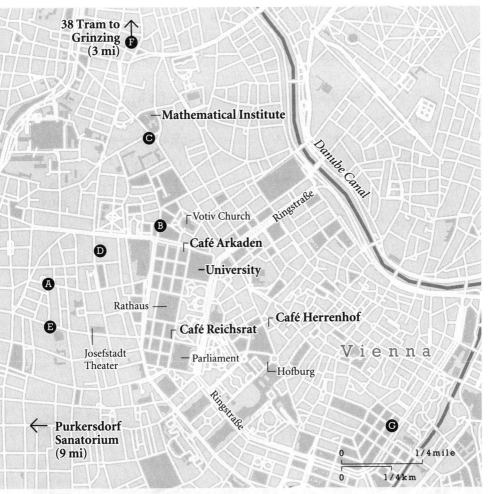

38 Tram to Grinzing (3 mi)

Mathematical Institute

Danube Canal

Votiv Church

Ringstraße

Café Arkaden

University

Rathaus

Café Herrenhof

Café Reichsrat

V i e n n a

Josefstadt Theater

Parliament

Hofburg

Purkersdorf Sanatorium (9 mi)

0   1/4 mile

0   1/4 km

Gödel's Residences

A Florianigasse 42 (October 1924–April 1927)

B Frankgasse 10 (April–July 1927)

C Währinger Straße 33 (October 1927–July 1928) and Café Josephinum

D Lange Gasse 72 (July 1928–November 1929)

E Josefstädter Straße 43 (November 1929–November 1937)

F Himmelstraße 43 (November 1937–November 1939)

G Hegelgasse 5 (November 1939–February 1940)

dominate the conservative and Catholic countryside, the new constitution made Vienna its own federal province. Within its boundaries the Social Democrats held a firm grip on power, and from that base launched a series of progressive reforms that would later be nostalgically remembered under the banner of "Red Vienna." Funded by a progressive housing and luxury tax, the city set to work building a socialist idyll: 60,000 units for workers' housing were created in vast apartment blocks outfitted complete with laundries, day care facilities, and lecture rooms; programs for public health and worker safety were vigorously promoted; and an exemplary adult education initiative, capped by the Volkshochschule—the "People's University"—brought literature, culture, and science to the masses, and employment to many of Gödel's colleagues in a generation for whom university posts would be exceedingly hard to come by.[5]

The coffeehouses where the mathematicians from the university liked to meet—the Café Arkaden, right by the university, with its outdoor tables in the summer comfortably spaced under the long arcaded gallery open to the sidewalk that gave it its name; the Café Reichsrat

Café Arkaden, near the university

and the venerable Café Central a few blocks farther on; and a few steps from the Central the new Café Herrenhof, with its Jugendstil furnishings, roof of bright yellow glass, windowed alcoves, and white marble tabletops which the mathematicians found convenient for writing equations upon, "spacious, light, sumptuous, impersonal" with the air of the new Red Vienna, in the words of the coffeehouse habitué Anton Kuh, where "the patron saint was Dr. Freud" and everyone with a political or revolutionary bent was to be found—once again buzzed with life, conversation, and intrigue. Among the regulars at the Herrenhof were the glittering literary clique that included the novelists Robert Musil, Joseph Roth, and Hermann Broch (the last to become Gödel's fellow mathematics student, when he returned to the University of Vienna in mid-life to take up his long-deferred boyhood passion for the subject), and just as often the small but devoted circle of young friends and intellectual sparring partners Gödel would grow close to over the next several years, all of whom were destined to join the ranks of the most famous mathematicians and philosophers of science of the twentieth century: Karl Menger, Rudolf Carnap, Olga Taussky, Herbert Feigl, and Alfred Tarski, among many others.[6]

If science was in the air of Brünn, it ran in the veins of Vienna of the 1920s. One thing the war and loss of empire had not destroyed was a passionate public interest in intellectual developments that rivaled the city's more well-known devotion to music, theater, and literature. When Albert Einstein came to town to give a public lecture on the Theory of Relativity in 1921, the organizers had to hastily change the planned venue from a six-hundred-seat lecture hall in the public education

Einstein in Vienna, 1921

institute, the Urania, to the Großer Konzerthaus, the largest concert hall in the city, with a capacity of two thousand. Tickets sold out in two days. With extra seats and standing room, the actual size of the audience as reported in the newspapers approached three thousand.[7]

When Karl Menger a few years later organized a series of popular lectures on "Crisis and Reconstruction in the Exact Sciences"—Menger would himself give the first popular exposition of Gödel's Incompleteness Theorem as part of the series—he had the same experience. Tickets, priced the same as a seat at the famed Vienna Opera, were snapped up instantly by a Viennese public equally eager to hear about such topics as "Is There an Infinity?" or "The Crisis of Intuition." As Menger said, "Many members of the legal, financial, and business world; publishers and journalists, physicians and engineers took intense interest in the work of scholars of various kinds. They created an intellectual atmosphere which, I have always felt, few cities enjoyed."[8]

Vienna in the 1920s was, too, the world capital of cranks, paranoids, megalomaniacs, and conspiracy theorists. The former monk turned sex and eugenics theorist Joseph Adolf Lanz garnered a considerable following with his pamphlets and books explaining the "racial value index" he had devised, warning of the attraction of "German women" to "primitive-sensual half-monkeys"; the Viennese engineer Hans Goldzier expounded his theory that all of science, including Newton's law of gravity, was false, and that humans were governed by electricity; and another Viennese engineer, Hanns Hörbiger—whose crackpot ideas greatly impressed Adolf Hitler—propounded a "glacial cosmogony" which asserted that "cosmic ice" was the basic ingredient of the universe, and that the "Nordic-Germanic" race had originated in a frozen Atlantis in the ice-cold north. Even the greatly talented Viennese man of letters Egon Friedell was taken with the idea that history runs in mystical 2,100-year-long cycles; and more than a few otherwise serious scientists—including Gödel's soon-to-be adviser, the eminent mathematician Hans Hahn—were ready to regard séances and other supposed parapsychological phenomena as worthy of legitimate scientific investigation.[9]

At the center of it all, the University of Vienna still enjoyed its unparalleled reputation as a world leader in philosophy, medicine, legal

"Philosophers' Staircase" in the main university building

theory, and mathematics. It was, in short, a heady time to be arriving as a student at a five-centuries-old university, just at the moment when relativity and quantum mechanics were revolutionizing physics, when the successors of Mach and Boltzmann were pushing back the boundaries of philosophy into the rigorous world of science, and when questions at the very heart of mathematics were succumbing to the confident prediction of the great German leader of the field, David Hilbert, that through "pure reason," *every* mathematical problem is solvable. "In mathematics," he proclaimed, "there is *nothing* unknowable!"[10]

"NOBODY CARED WHAT YOU DID DURING THOSE EIGHT SEMESTERS"

For many a newly arrived student, the abrupt transition from complete supervision of their education to equally complete freedom was more than a bit of a shock. At the university there were no prescribed courses, no graded homework assignments, no examinations, and until one had established oneself almost no direct contact with professors, and thus no one to offer much in the way even of guidance. "When you entered the university," recalled Olga Taussky, "you were given two little books. The smaller one was an identity card with your picture in it and a student had to carry it all the time. . . . The second one recorded the courses you had registered for. There was a minimum of hours you had to take. However, nobody bothered whether you attended these courses." The only degree offered in the Faculty of Philosophy, which incorporated all of the liberal arts and sciences, was the doctorate; after registering for eight semesters a student could present a thesis and, if it were deemed worthy, ask to be examined by two professors from one's major field, one from a minor subject, and two from philosophy. "In principle," Taussky said, "nobody cared what you did during the eight semesters."[11]

Taussky, a year behind Gödel, had met him shortly after arriving at the university in the fall of 1925. They were both enrolled in a seminar, Introduction to Philosophy of Mathematics, which involved reading and

discussing a German translation of Bertrand Russell's recently published *Introduction to Mathematical Philosophy.* The professor was the forty-three-year-old Moritz Schlick, who was to make a lasting mark on the lives and thought of Gödel's generation in Vienna.

Taussky was, by her own account, a "serious, hardworking" as well as "very worn-out, grief-stricken person." Born into a Jewish family in the Moravian town of Olmütz, forty miles northeast of Brünn, she had always been encouraged by her father to pursue education and to do well in school. An industrial

Olga Taussky

chemist with a love of writing and learning, her father had hoped his three daughters would do something in the arts, though not necessarily as a career. Olga fell in love with math in her early teens, feeling from the start a deep sense of its romance and creativity. In her final year of high school her father died suddenly, leaving the family with some savings but no income. Her mother thought the best thing would be for her to join her older sister in taking over her father's consulting business, which both daughters had shown an aptitude for. But that summer she had one day spoken to a family friend, a much older woman who revealed that she too had once dreamed of studying mathematics. "That was more than I could take," Taussky remembered. The picture of herself years in the future regretfully speaking the same words to a younger woman was "unbearable," and she determined to follow her own path.[12]

The second week of the seminar, Gödel noticed his intent younger classmate with the head of short dark curls when she summoned the courage to ask a question about the connection between the axioms of geometry and of number theory. When Schlick asked for someone to report on the day's discussion at the start of the next seminar, Gödel after a pause volunteered. The following week he began his report, "Last week somebody asked the following question. . . ." But he soon learned her name, and the two became fast friends. Though "he seemed a rather silent man," Taussky remembered, he was always willing to assist others with their mathematical ideas and his help "was much in demand." It became "quite natural to me to phone him at his residence and have a little chat," she said, and even years later, after the war, when they saw each other again in Princeton after having been separated by careers and circumstances for a decade, "we renewed our friendship instantaneously."[13]

Gödel intended to study physics, and in his first two years he enrolled in a staggering thirty-plus hours a week of lectures, among them optics, electrical theory, theoretical physics, mechanics of deformable bodies, experimental physics, partial differential equations for physicists, kinetic theory of matter, and relativity theory.[14]

Years later he mentioned two lecturers who seduced him away from physics to the more abstract realms of pure mathematics.[15] Heinrich Gomperz, who taught an introductory survey course on the major problems in philosophy, was legendary for having played a pivotal part years earlier in bringing Ernst Mach to Vienna. His father, Theodor Gomperz—another Jewish native of Brünn—was professor of classical philology at Vienna, having spurned a career in his family banking business to become one of the world's foremost authorities in Greek philosophy. Heinrich, captivated by reading a lecture of Mach's, urged his father to support the appointment of the physicist to the new philosophy chair, and forever after revered Mach as "the incarnation of the scientific spirit."[16]

But the professor who truly tipped the scales for Gödel was the charismatic Philipp Furtwängler, a self-taught number theorist whose lectures drew as many as four hundred students, more than there were seats in the hall. A cousin of the famous orchestral conductor, he had a flair for showmanship as a lecturer, heightened by the drama of a paralysis that had left him unable to walk without a cane and the support of two assistants. Once seated in a chair on the stage, he would deliver a flawlessly prepared lecture without using any notes, while an assistant wrote equations on the blackboard for him— a task which Olga Taussky described as extremely challenging when she later sometimes took on the job.[17]

Philipp Furtwängler

Number theory is the branch of mathematics that deals with its most basic building blocks, the whole numbers 1, 2, 3, . . . and the patterns and relationships that appear among them. In asking fundamental questions about the behavior of numbers—What determines if a number that is itself a square, such as 4 or 9 or 25, is the sum of two other squares? Are there infinitely many prime numbers of the form $N^2 + 1$? Can every whole number greater than 2 be written as the sum of two prime numbers?—number theory is an exploration of the nuts and bolts of mathematics itself.

Even while still planning to concentrate in physics, Gödel quickly began plunging more deeply into mathematics and philosophy, both in the lectures he enrolled in and in his own extensive independent reading. In a self-excoriating list of his faults he set down in a diary of his conversations and

In his student days

thoughts that he kept in Gabelsberger shorthand during the fateful year of 1937–38, Gödel berated himself for doing "nothing thoroughly":

> You do everything with haste (lack of time)—especially in acquiring knowledge (learning) You should make note of the propositions you want to strictly formulate and decide for yourself (in a methodical fashion) which ones you want to remember. . . .
> Comment: Dependence on outside value judgment (signs of esteem etc.) is partly due to the fact that one is not secure in one's own value judgment[18]

Such self-judgment spoke more to his perfectionism than any actual lack of diligence. Gödel would astonish colleagues throughout his life with his insights into whatever questions he turned his mind to: abstruse mathematical problems fellow students were struggling with, his highly original explorations of Einstein's general relativity equations, solutions to fundamental unresolved problems of set theory. Even in his more casual forays into areas that piqued his curiosity—economic theory, statistical patterns of fog formation, the philosophical writings of Hegel, the innumerable odd geographic and political facts he sedulously committed to memory while perusing the *World Almanac*—no one would have accused him of a lack of thoroughness.[19]

In fact, ever since his self-study of calculus and Kant as an impatient high school teenager, he had learned to have confidence in his ability to tackle subjects on his own. In his first semester at the university he began requesting books on the subjects that were now drawing him away from the more practical worlds of physical science to the ethereal realms of pure thought. He read the classics of mathematics by Euclid, Euler, and Lagrange, advanced works on topics such as partial differential equations, and Kant's *Metaphysical Foundations of Natural Science*.[20]

With practicality in mind, Gödel also took courses needed to prepare for the *Lehramtsprüfung*—the arduous examinations, in some ways even more of an ordeal than those required for a doctorate, that one

Hans Hahn

had to pass to become a schoolteacher at a Gymnasium or Realgymnasium in Austria. In that era of economic austerity, teaching high school would be the enduring lot of more than one of Gödel's gifted contemporaries: a considerable number of schoolteachers in Austria between the wars were otherwise unemployable PhDs. During his first six semesters Gödel dutifully enrolled in a series of mandatory lab courses for physics teacher candidates along with lectures in the history and theory of education, teaching of mathematics, child psychology, even school hygiene. He would stick with this backup program, hedging his bet even after he made the fateful decision to change his major from physics to mathematics.

He made the decisive switch to mathematics at the start of the 1926–27 academic year. Gödel's thesis advisor, Hans Hahn, was not only a versatile mathematician, but the animating force behind the remarkable discussion group that for a brief, electrifying decade would make Vienna the most exciting place in the world for new philosophical thought. The sparking clash of original and young thinkers in the circle that formed around Hahn would break down the walls between philosophy and science, bring the brash and mesmerizing ideas of Ludwig Wittgenstein to wider notice, and form the milieu out of which Karl Popper's great clarifying insights on falsifiability and Kurt Gödel's on unprovability emerged. "Tales of murder and suicide, of love affairs and nervous breakdowns, of political persecution and hair's-breadth escapes all have their place in the rich tapestry of the Vienna Circle," wrote the mathematician and historian Karl Sigmund, "but the tapestry's main thread is the unbroken stream of heated debates among its members."[21] Few crucibles of creative energy have been so productive.

Since his student days Hahn had nursed a longing to add the serious study of philosophy to his many-sided interest in mathematics. "I am not easily affected by emotions," he wrote in 1909 to one of his old university friends, the physicist Paul Ehrenfest, then in far-off St. Petersburg, "but to a friend who is as far away as you are, I must confess: At times, in my fleeting attempts to dip into the metaphysics of Aristotle, I have felt awestruck—and I deeply regret the lack of any opportunity to ponder these things in depth."[22]

Hahn was writing from a far-off outpost himself, the University of Czernowitz, his first academic position after completing his degree and postdoctoral work at the University of Vienna. Czernowitz, the provincial capital of Bukovina, on the Russian frontier at the eastern end of the empire, was a twenty-six-hour train ride from Vienna. At a farewell gathering of his friends at a Viennese coffeehouse, he confidently predicted his future triumphant return as a professor at the University of Vienna, when they would be able to resume their discussions but this time with a real *Universitätsphilosoph* to help guide their circle of *Kaffeehausphilosophen* in clarifying their thoughts.[23]

A dozen years later Hahn made good on the first part of his boast, winning a chair in mathematics at his alma mater. The following year, in 1922, he was able to complete his promise by using his influence to bring to Vienna, to fill a vacant chair in philosophy, the *Universitätsphilosoph* he had been seeking: Moritz Schlick, a close associate of Einstein and a true heir of Ernst Mach's empirical tradition. Hahn, now at last in possession of the opportunity to ponder these things in depth, saw the goal of the group's philosophical explorations as nothing short of a total overthrow of the "world-denying" philosophy that had dominated German thinkers, with their abstract concepts and invented terminology. The only knowledge of the real world came from experience. "Any knowledge gained by pure thought," he insisted, "appears to me to be completely mystical."[24]

Shortly after returning to Vienna, Hahn began an intensive study of symbolic logic, which he saw as offering a way of placing philosophy on the rigorous footing of logic and clearing away all of the wooly imprecision that had plagued the field for generations. During his first semester Gödel enrolled in the seminar Hahn led on Bertrand Russell's magnum opus of mathematical logic, *Principia Mathematica*. Russell and his coauthor, Alfred North Whitehead, had set out to show that all of mathematics could be derived from the propositions of logic. It was a sign of Hahn's philosophical iconoclasm that he not only recognized such a work as being about philosophy at all, but on its basis was willing to venture to declare in print that one day Russell might well be regarded as the most important philosopher of his time—at a time when few European philosophers had ever read Russell's work.[25]

Hahn also shared with Russell strong left-wing and pacifist views. Serving in the Austrian Army during the war, Hahn had been severely wounded on the Italian front in 1915 when a bullet tore through his lung. By then Czernowitz was in Russian hands, so he had no job left to return to. In 1917 he received an appointment at the University of Bonn, where he immediately made himself persona non grata by handing out antiwar leaflets. At the same time in England, Russell was serving a six-month sentence in Brixton Prison for his antiwar agitation, using what he later described as an "in many ways quite agreeable" period of enforced and uninterrupted solitude to write his *Introduction to Mathematical Philosophy*.[26]

Hahn epitomized the air of scientific optimism that filled mathematics and philosophy at the start of the twentieth century. Karl Menger, who completed his doctorate under Hahn in 1924, described him as "a strong, extroverted, highly articulate person who always spoke with a loud voice." Karl Popper, who would forever be grateful for the crystalline clarity of Hahn's calculus lectures, which at last allowed him to master that subject, recalled that "the personal impression of Hahn was that of a quite uncommonly disciplined person."[27] His lectures, like Furtwängler's, were meticulously prepared: "He applied a technique that I have never seen anyone else carry to such extremes," Menger said. "He proceeded by almost imperceptible steps and at the end of each hour left his audience amazed at the mass of material covered." He also had an incomparable talent for attracting talent. His "penetrating criticism, the clarity of his ideas, and his skill in presenting them," in Menger's words, would provide a launching pad for many of the brilliant young thinkers he encouraged.[28]

His student Kurt Gödel struck others as having some of the same qualities: the word "precision" appears often in their recollections of the way he would go to the very heart of a problem, reducing confusion to order with a few deft insights. It was that same "interest in precision," he told his friend the philosopher Hao Wang many years later, that led him to make the move from physics to math in 1926. In 1949, Oskar Morgenstern, noting the widespread surprise that had greeted Gödel's recent work on relativity, wrote in his diary: "Everyone is amazed at his deep knowledge of physics. They did not know that he originally studied physics & turned away from it because it was logically so messy to him. This he told me years ago." It was an early sign of his inclination to turn his back on the Viennese empiricism of Mach, Hahn, and Schlick in the deeper conviction that there were truths to be found not just in the empirically perceivable, but perhaps more beautiful and enduring ones in the realm of abstract conceptions, where they awaited human discovery not through tangible perception, but by thought alone.[29]

In an especially revealing entry in one of his private shorthand notebooks, Gödel reflected on his abiding interest in getting to the very root of things, in science and in life:

> I am apparently neither talented nor interested in combinatorial thinking (card games and chess, and poor memory). I am apparently talented and interested in conceptual thinking. I am always interested only in how it works (and not in the actual execution). Therefore, I should dedicate myself to the foundations of the sciences (and philosophy). This means: Not only the foundations of physics, biology and mathematics, but also sociology, psychology, history (world, earth, history of mankind). . . . I was originally interested in explaining the phenomena of everyday life in terms of higher concepts and general regularities, hence physics.

"So he switched to logic," Morgenstern added in an ironic footnote, "which he also did not find in order!" But in a world that would often fill Gödel with anxieties, the realm of pure mathematical logic beckoned with the enfolding reassurance of certainty so noticeably absent everywhere else, except only perhaps in the atavistic feelings of childhood security that he would at times desperately cling to throughout his life. Writing to his mother years later, when he was fifty-nine and rereading the fairy tales he had loved in his youth, he wistfully observed, "You are right that fairy tales are like oases among the other arts. For only they depict the world as it should be and as if it had meaning, whereas in tragedy the hero is killed off and in comedy the ludicrous (therefore also something bad) is emphasized."[30]

HILBERT'S CHALLENGE

Olga Taussky was not alone in her time and place in seeing in mathematics a romantic calling. Hermann Broch's first poem, written while he was still unhappily playing the dutiful role of heir to his family's textile fortune, was titled "Mathematisches Mysterium," and it tried to capture his awe at the magical way mathematics connected with truth:

Built of nothing but solitary thought,
A structure rises precipitously upward:
And joins the clustering stars,
Illuminated by a distant divinity.

As in Robert Musil's *The Man Without Qualities*, the hero of Broch's 1933 novel *The Unknown Quantity*, written after Broch had spent five years in his mid-forties studying mathematics and philosophy (during which time he took many of the same classes as Kurt Gödel), is a mathematician. Challenged by a friend to explain what he believes in, Broch's hero tries to describe the awesome beauty and power of mathematics, likening it to the complexity of a weaving: "What was his conception of mathematics? A bright network of shining reality spread out infinitely, and one had to feel one's way from knot to knot; yes, it was something like that, a complicated web of cosmic weave, like the world itself, which one had to unravel in order to get hold of reality."[31]

And in a passage that sounds much more like Kurt Gödel than his empiricist mentors who relentlessly sought truth in nothing but physical experience, Broch describes mathematics as a parallel, unseen world that is nonetheless as real to him as God is to the believer:

For the world of mathematics in which he moved, with all its algebraic symbols, its theoretical interrelation of sets of numbers, its infinitesimal infinitude in small things as in great, found only the crudest expression in the world of concrete fact; and even the delicate constructions of physical science, evolved from intricate and ingenious experiments, even the calculability of these physical phenomena, formed in sum only a small, inadequate, pale reflection of the manifold thought-complexity of mathematics, which was embedded in the concrete visible world as an original principle, far

beyond the concrete, spanning the whole universe and yet imma-
nent in the reality of the universe as in its own reality.[32]

David Hilbert, who reigned as the undisputed king of mathematics in the German-speaking world, unabashedly shared that romantic
vision. He liked to tell of a former student who had given up the subject. "That Schmidt, he didn't have enough imagination for mathematics," Hilbert disdainfully observed. "Now he has become a poet. For
that he had just enough."[33]

In 1900, at the second International Congress of Mathematicians,
held that year in Paris, Hilbert had delivered a famous address that
could have been a model for Broch's mathematical protagonist, with
the lyrical and poetical view of their profession he expressed, and its
promise he foresaw in the coming century.

> Who of us would not be glad to lift the veil behind which the future
> lies hidden; to cast a glance at the next advances of our science and
> at the secrets of its development during future centuries! . . .
>
> However unapproachable these problems may seem to us and
> however helpless we stand before them, we have, nevertheless, the
> firm conviction that their solution must follow by a finite number
> of purely logical processes. . . . This conviction of the solvability of
> every mathematical problem is a powerful incentive to the worker.
> We hear within us the perpetual call: There is the problem. Seek its
> solution. You can find it by pure reason, for in mathematics there is
> no *ignorabimus*, nothing unknowable! . . .
>
> The organic unity of mathematics is inherent in the nature of
> this science, for mathematics is the foundation of all exact knowl-
> edge of natural phenomena. That it may completely fulfill this high
> mission, may the new century bring it gifted masters and many
> zealous and enthusiastic disciples.

Hilbert then listed ten important unsolved problems in mathematics
which he challenged his listeners to tackle, subsequently expanding
the list to twenty-three when the speech was published the following

year. He admonished them, too, to look upon mathematics as a human endeavor like art or music that did not belong only to specialists in their field. "A mathematical theory," he told the assembled mathematicians, "is not to be considered complete until you have made it so clear that you can explain it to the first man whom you meet on the street."[34]

The everlasting beauty of mathematics, he concluded, was that it was *inexhaustible*. Every new solution is the seed of new ideas and new problems to be solved.

For forty years Hilbert made Göttingen a world capital of mathematics, training a large cadre of students and issuing inspiring challenges to his colleagues in ringing language that imbued their work with an aura of heroic endeavor. He combined those gifts with a personal joie de vivre and élan that worked equally to infuse their profession with panache, the exact opposite of the image of the mathematician as emotionless loner. Although he suffered black periods that occasionally debilitated him, that was the price most great mathematicians paid, his colleague Richard Courant observed. "Almost every great scientist I have known has been subject to such deep depressions," Courant wrote in reference to Hilbert's periodic struggles against despair. "There are periods in the life of a productive person when he appears to himself and perhaps actually is losing his powers. This comes as a great shock."[35]

But optimism ran through Hilbert's core, as did his delight in puncturing solemnity. He wore a very unprofessorial jaunty straw hat and short sleeves in summer, loved dancing and parties where he could find a pretty young girl to explain mathematics to, at age forty-five took up bicycle riding for the first time, and whenever he could, worked outdoors in his garden, using a large blackboard hung on the neighbor's wall. "If you don't see the professor," his housekeeper would tell visitors as she directed them into the garden, "look up in the trees."[36]

Hilbert never lost the sharp Königsberg accent of his birthplace in the easternmost outpost of Prussia, nor the spirit of Kant that still filled that city a century after his death, and whose tombstone inscription he often saw while growing up: "Two things fill the mind with ever new and increasing admiration and awe, the starry heavens above me and the moral law within me."[37]

In September 1928, Gödel's mentors Hans Hahn and Karl Menger traveled to Bologna for the quadrennial Congress of Mathematicians. It was an occasion of considerable moment on several scores. Due to the war, no Congress was held in 1916, and under the influence of French leaders of the organization, still brimming with hatred of their ex-enemies, the next two meetings in 1920 and 1924 banned German and Austrian mathematicians from participating. The 1928 Congress was the first in which they were welcomed back.

A number of Germans angrily demanded that their countrymen refuse to attend, in protest of their previous treatment. But Hilbert, defying their antipathy, personally led a delegation of sixty-seven mathematicians from Germany to the Bologna Congress. In a show of reconciliation, Hilbert had been invited to give the opening address, just as he had twenty-eight years earlier in Paris. "Mathematics," he wrote in his prepared remarks, "knows no races. . . . For mathematics, the whole cultural world is a single country." When Hilbert entered the hall at the head of the German delegation, there was a moment of total silence, followed by an outburst of sustained applause as the entire audience spontaneously rose to its feet.[38]

One of the most vociferous in calling for a boycott was the Dutch mathematician L. E. J. Brouwer, who had already become a thorn in Hilbert's side owing to their fundamental disagreements over the conception of mathematical proof. Brouwer was also a committed Aryan nationalist who complained that too many papers by "Ostjuden," Jews from Eastern Europe, were being published in the prestigious *Mathematische Annalen*; after the Second World War he would be temporarily suspended from his university position as punishment for collaborating with the Nazis during their occupation of Holland. Hilbert's removal of Brouwer from the editorial board of the *Mathematische Annalen* in 1928 further embittered feelings between the two leaders of the field.[39]

The Italian government for its part saw the Congress as an opportunity to showcase the new Fascist state. Il Duce himself lent his name as president of the honorary committee, and the mayor of Bologna grandly announced in his welcoming speech, "The Fascist Bologna is proud to offer its hospitality and to exhibit what Bologna has become under the vivifying impulse of Fascism."[40]

Hilbert, true to form, presented to the assembled mathematicians a new series of challenges. All went to the heart of the *Grundlagenkrise*, the foundational crisis, currently roiling mathematics. In his address entitled "Probleme der Grundlegung der Mathematik"—"Problems of Laying Foundations for Mathematics"—Hilbert now posed four problems whose solution he believed would at last place all of mathematics on an unshakeable, rigorous footing.

The first two involved proving the *consistency* of mathematical systems: that they contain no contradictions. The third was to prove the parallel property of *completeness*: that every valid statement within a system can be derived from its basic axioms. The fourth and last was to prove the completeness of the fundamental system of logic known as first-order, or predicate, logic. The aim, Hilbert explained, was to remove "once and for all" any question about the soundness of the systems mathematicians depended on to produce new results. He looked, he said, to "the devoted cooperation of the younger generation of mathematicians," which would be needed to achieve a full solution of this monumental task.[41]

The twenty-two-year-old Gödel wasted no time answering the call. Within six months he would, in his PhD thesis, solve Hilbert's fourth problem. The following year, even more astonishingly, he was to prove the impossibility of anyone's ever solving the first three, in the work that would secure his enduring fame by upending the very idea that any mathematical system could be both consistent and complete.

THE BEAR'S DEN

For most members of the Faculty of Philosophy at the University of Vienna, Hans Hahn's declaration of war against world-denying philosophy was not the herald of a new age but a threat to their livelihood, one they were couching increasingly in terms of the political and anti-Semitic dividing lines that were fracturing Austria in the 1920s. Hahn, Schlick, and their growing circle of protégés were, in Karl Sigmund's concise description, a left-wing pocket in a right-wing university of a left-wing town in a right-wing country. As an outspoken socialist and a Jew, Hahn was a magnet for attacks by conservatives, but the reactionary thinkers who dominated the faculty were equally quick to brand *any* ideas that challenged their conservative values or traditional avenues of thought as "Jewish" or "un-German." Positivism, psychoanalysis, marginal economic theory all came under attack as examples of the "individualistic" and antinational ideas of "Jewish science" that threatened the romantic "holism" preached by the old guard.[42]

The ultranationalist economist Othmar Spann, who now held sway in the Faculty of Law and Political Science where the elder Carl Menger had developed his revolutionary insights into economic behavior a generation before, denounced as un-German any such studies focused on individual needs and wants. The only proper work for philosophers, according to the champions of "Christian worldview" who likewise dominated that area at the university, was the *history* of philosophy.[43] A guidebook to Vienna published in 1927 in the irreverent series *Was nicht im Baedeker steht* ("What's Not Found in Baedeker") included under the heading "Peculiarities that one must get used to" the advice not to do anything too "interesting or original, otherwise you will suddenly, behind your back, become a Jew."[44]

The universities had long been a hotbed of German nationalism and anti-Semitism: indeed, as the historian Bruce F. Pauley observes, "it was the Austrian universities that helped to make anti-Semitism respectable throughout the country." Since the 1870s, fraternities that venerated Germanness had been a vocal and often violent part of student life at the university. Devoted to the glorification of dueling, drinking, and militarism, and to the denigration of parliamentary democracy, liberalism, and laissez-faire capitalism, they were among the most enthusiastic supporters of the new "scientific" anti-Semitic theories that ascribed a host of loathsome traits to Jews' inherent racial characteristics.[45]

By the 1920s the middle-class insecurities that inflamed Austrian anti-Semitism were concentrated at the universities; many sons of small tradesmen and artisans who enrolled were the first in their families to seek a university education, hoping to qualify for government service or the professions. Those insecurities were heightened by government

Students fleeing anti-Semitic attack on the university's anatomy lab

austerity measures, high unemployment, overcrowding of the universities, and the particularly acute competition for academic positions. The sociologist Max Weber had warned in a speech back in 1917 that it was a "mad hazard" for anyone to seek an academic career in science. As for Jewish students, he said, the only honest advice were the words posted at the gates to Dante's hell: *lasciate ogni speranza*, abandon all hope.[46]

The situation had only grown worse in the years since. Several powerful organizations at the University of Vienna began intensifying their intimidation of Jewish students and faculty. In 1923 the German Students' Union demanded that all books in the library written by Jews be marked with the Star of David. Vienna's *Neue Freie Presse* reported that on a stroll around the main building of the university, a visitor would encounter little but anti-Semitic posters and hate literature, such as the notorious fabrication *Protocols of the Elders of Zion*, on sale with the university rector's full approval. The German Students' Union began drawing up blacklists of "undesirable" professors and handing out leaflets warning fellow students not to attend their lectures. Noting with disgust the "increasing Jewifying [*Verjudung*] of the university in the spirit of cursed liberalism," a statement on "Race and Science" issued by the Union named two hundred Jewish and liberal professors to be boycotted, among them Sigmund Freud, the legal scholar and author of the Austrian Republic's constitution Hans Kelsen, Karl Menger, and Moritz Schlick.[47]

Anti-Jewish riots, directed particularly against the students and laboratories of the medical school with its heavily Jewish enrollment, had been a sport of the German fraternities going back to 1875, but these now became much more violent. A favorite target was the anatomy institute of Julius Tandler, who had been a prominent architect of the public health programs of Red Vienna. In 1927, an attack by German-nationalist students seriously injured several students and severely damaged classrooms, marking the first in a series of steadily escalating assaults, culminating in attacks with brass knuckles, whips, knives, and iron rods.[48]

Though the violence led to the university being repeatedly shut down, it was met with only half-hearted condemnation by university administrators, who tacitly or even openly shared the students' anti-Semitic enthusiasms. An alliance of Catholic and German-nationalist academics and administrators, the Deutsche Gemeinschaft, had pledged itself to fight "anarchic tendencies"—a not very subtle codeword for liberals and Jews. From 1920 on the Minister of Education was always a clerical–conservative, a German-nationalist, or even a pro-Nazi, and the university's rector, chosen in rotation each year from the schools' faculties, reliably reflected that official outlook. During his year as rector, Wenzel Gleispach, a jurist and later Nazi Party member, instituted a *numerus clausus* limiting the number of Jewish students; when the Constitutional Court of Austria overturned the rule, German-nationalist students responded with yet another wave of violent riots.[49]

Operating even more effectively behind the scenes was a small network of prominent members of the Faculty of Philosophy who worked simply to block any appointments or academic promotion for Jewish and left-wing teachers. Led by Othenio Abel, a paleobotanist, the group of nineteen professors met in a small seminar room that housed the biology department's collection of animal skeletons and taxider-

The "Bärenhöhle"

mic specimens, which gave the clandestine committee its nickname: the *Bärenhöhle*, or Bear's Den. In a letter to a friend, a Catholic priest, Abel boasted, "I really take the credit for binding our anti-Semitic group together so closely that we form a strong phalanx. And while I have lost and still lose much time and energy, I hold on to the knowledge that this work is perhaps as important as making books."[50]

The "phalanx" had tried to scuttle the appointment of Moritz Schlick by demanding an investigation to determine if he was Jewish (he was not). They were more successful in blocking later appointments of any of his protégés and in sabotaging approval of theses by Jewish students. Edgar Zilsel, a student of mathematics, philosophy, and physics whom Gödel would get to know well through the discussions in Hahn's and Schlick's philosophical circle, was an early victim. Zilsel's *Habilitation*, the postdoctoral thesis that conferred the right to lecture at a university, was rejected on the grounds that it was "too one-sidedly rationalistic" to constitute an acceptable work of philosophy.[51]

Always wary of revealing his personal thoughts to others on non-scientific matters, Gödel rarely spoke of his political opinions, keeping his own left-of-center views mostly to himself and scarcely commenting even about the upheavals taking place at the university. Yet his distance and detachment from overt political activity would prove no refuge amid the rising tide of political polarization and violence over the coming decade. His mere association with progressive and Jewish thinkers was enough to place him in peril in an Austria where ideological opponents were coming to be seen as mortal enemies to be eliminated.

The dangerous fissures within the university mirrored those of Austrian politics at the national level, where both the Social Democrats and Christian–Socials were increasingly contemplating a resort to force to settle their implacably diverging views of Austria's future. Both parties had formed paramilitary "self-defense" organizations, and it was only a matter of time before a serious armed clash took place. The spark

Justizpalast attack, 1927

came shortly after the German students' riots in June 1927, which again temporarily closed the university. In a small village east of Vienna, a group of right-wing Heimwehr militia fired into a meeting of the socialist Schutzbund and its supporters, killing a child and a disabled war veteran. At a trial in Vienna on July 14, a jury acquitted the three men accused of the murders. The next day mobs of workers poured into the streets. The city's Social Democratic leaders quickly lost control of the situation, unable to contain the anger of their members, who set fire to the Palace of Justice. Determined to make an example, the conservative national government issued rifles and machine guns to the police and authorized them to use deadly force to "restore order." The crowd around the burning building was dispersing when police arrived and opened fire. When the shooting was over, eighty-nine were dead and six hundred severely wounded.[52]

Gödel and his friends Marcel Natkin and Herbert Feigl were briefly trapped by the fighting; Natkin wrote to Gödel a few days later trying to make light of the gravity of the situation: "Imagine, that Feigl could not get away on Friday and had to wait for the end of the 'Revolution.' Hopefully you got home okay."[53]

But in retrospect they all would see that brief but bloody paroxysm as the end of Austrian democracy. "Before 15 July there was a chance, however slight, that Austria's Left and Right might find a formula for coexistence," wrote George Clare in his memoir of the lost Vienna of his youth. "After it there was no hope."[54]

· 4 ·

Floating in Midair

CITY OF CIRCLES

Some time around 1926 Gödel was invited by Hans Hahn to join his rarified philosophical discussion circle. It was a signal honor: its meetings, held every other Thursday evening at six o'clock, were limited to a select group that never exceeded ten or twenty. Karl Popper, who

A serene Moritz Schlick

would later become one of the most influential philosophers of science of the twentieth century, never was asked, and still felt the sting a half century later.[1]

The circle had been inaugurated two years earlier when two of Moritz Schlick's favored students—Friedrich Waismann and Gödel's soon to be fast friend Herbert Feigl—urged Schlick to launch a discussion group that would pick up where Hahn's "prehistoric" circle, as they humorously called it, had left off. Hahn's group of *Kaffeehausphilosophen* had included among its more notable participants the physicist Philipp Frank and the applied mathematician Richard von Mises, both of whom would finish their careers at Harvard, and Otto Neurath, who defied easy description altogether.[2]

Neurath, a large, energetic man with a bushy red beard and equally untamed enthusiasms, would, along with Hahn and Schlick, be a principal driving force of the new circle. A founder of the study of war economics, Neurath had served several months in a German prison in 1919 for his part in the short-lived Soviet government of Bavaria, where he worked to formulate a master plan for the economy. He was married to the sister of his old schoolfriend Hans Hahn, Olga, who had been one of the first women to receive a doctorate in mathematics from the University of Vienna, despite losing her eyesight at age twenty-two; and Neurath, along with all of his other endeavors, devoted countless hours to helping her continue her work in mathematical logic. He was full of ideas for projects to upend society, educate the masses, and drive a final stake through the heart of outdated metaphysical thinking. "Some kind of a professional warhorse," Robert Musil described him in his diary after meeting him on his release from prison. "But with explosive energy."[3] It was from Neurath that Gödel may have acquired the distinctive habit of filling a notebook with questions to raise with his colleagues, which he would then solemnly cross off once he had received a satisfactory answer.

Discussion circles dedicated to every imaginable intellectual topic had been the rage in Vienna since before the First World War, many of them devoted to various brands of philosophical investigation. There were circles on Kant, on Kierkegaard, on Tolstoy, on phenomenology, on the philosophy of religion. But the new group was distinctive for its strongly analytical cast. "All members of the Circle had a background of scientific research," said Karl Menger, "took the scientific method seriously, and in fact expected to obtain a consistent *Weltbild*—picture of the world—through what in the Circle was called *die wissenschaftliche Weltauffassung*," the scientific worldview. The group was distinctive, too, for being one of the few places in the still very hierarchical world of Austrian higher education in which members of three generations, men and women, Austrians and foreign visitors, met and debated on equal terms.[4]

Café Josephinum, which Gödel lived above in 1927–28

By the time Gödel joined the circle, Schlick had begun leading the discussions, which took place in what Menger called "a rather dingy room" next to Schlick's office on the ground floor of the Mathematics and Physics Institute. The building, completed just before the war on the newly named Boltzmanngasse, was a half mile from the main university building, near the famous Strudlhof staircase. For one year, Kurt and Rudi Gödel lived just across the street, at Währinger Straße 33, the ground floor of which housed the Café Josephinum where the group's informal discussions often continued.

Schlick led the meetings with a dignified informality. Menger recalled:

Those who arrived first at the meeting of the Circle would shove some tables and chairs away from the blackboard, which most speakers used. In the space thus gained they arranged chairs informally in a semicircle in front of the blackboard, leaving one long table for those who brought books or wished to smoke or take notes.

People would stand in informal groups until Schlick clapped his hands. Then the conversations stopped, everyone took a seat, and Schlick, who usually sat at one end of the table near the blackboard, announced the topic of the paper or the report or the discussion of the evening.[5]

In Menger's description, Schlick was an "extremely refined, somewhat introverted man," "very sincere and unassuming," "perfectly self-assured," yet always willing to learn and reassess his own views. "Empty phrases from his lips or the slightest trace of pompousness were unthinkable." Schlick had earned Einstein's respect for his essays and popular books explaining relativity and its philosophical implications. Shortly after his arrival at the university he wrote Einstein, "In Vienna there is much philosophizing going on now. Soon I hope to be able to present you some samples, which you will surely be interested in."[6]

Karl Menger, visiting Austin, Texas, in 1931

The "Schlick Circle" would be the basis of many of Gödel's most enduring friendships. He came to know his lifelong friends Feigl and Natkin through its meetings, and likewise Menger, with whom he enjoyed a warm friendship that would last until Menger's dismay at what he felt was Gödel's stunning naiveté over the situation in Vienna following the Nazi takeover led to a permanent cooling in their connection.[7]

The two had first met in 1927, soon after Menger's return to Vienna from Amsterdam, where he had gone on a Rockefeller Foundation fellowship to study with Brouwer. It had proved a tempestuous three years, culminating in a bitter priority fight over a discovery in topology claimed by both Menger and a former protégé of Brouwer's, Pavel Uryshon, whose life had been tragically cut short in a fatal swimming accident at age twenty-six.

Menger had been a precocious schoolboy, with dreams of literary fame. A classmate was the son of the famed Viennese playwright and litterateur Arthur Schnitzler, and at age seventeen Menger brashly sent Schnitzler a play he had written about Pope Joan, the apocryphal woman pope who had disguised herself as a man. Schnitzler recorded in his diary: "Talented fellow, but not literary. . . . His true calling: physicist. In reply to my question about long-term plans: 'I would most like to kill myself.' Certainly a very talented but possibly not quite normal young man." He added later: "May be a genius—but with megalomaniac and eccentric traits."[8]

Menger certainly had no shortage of self-confidence. After hearing the first lecture of Hahn's course on "New Developments in the Concept of Curves," he spent a feverish weekend working out a solution to the definition of dimension. A few days later he marched into his professor's office to present his result. Students did not normally presume to intrude on their professors this way. But Hahn, "who hardly looked up from the book he was reading when I entered," Menger recounted, "became more and more attentive as I went on." It was to become the subject of his thesis, and a major contribution to mathematics.[9]

"Shortly after assuming my position at the University of Vienna in the fall of 1927," Menger recalled, "I offered a quite well attended one-semester course which was on dimension theory. The name of one of the students who had enrolled was Kurt Gödel."

He was a slim, unusually quiet young man. I do not recall speaking with him at that time.

Later I saw him again in the Schlick Kreis. . . . I never heard Gödel speak in these meetings or participate in the discussions; but he evinced interest by slight motions of the head indicating agreement, skepticism or disagreement.[10]

Gödel and Menger shared a certain distance from the philosophical views of the group, particularly that of the more radical exponents of logical empiricism such as Hahn, Neurath, and the recently arrived German philosopher and logician Rudolf Carnap. "I was not a particularly active member of the Schlick Circle," Gödel told his mother after the war when she wondered why he was not mentioned in a newspaper article about Schlick and his followers, "and in some regard even in direct opposition to the predominant views there." Gödel could never reconcile himself to the positivist standpoint that knowledge derives solely from empirical observations of natural phenomena. Mathematical objects and a priori truth were as real to him as anything the senses could directly perceive.[11]

Rudolf Carnap

In Carnap's view, which he would later call "physicalism," philosophical assertions about ethics, values, and aesthetics were "pseudo-propositions." They were not even wrong, but literally meaningless. Carnap, who would more than any carry on the ideas of the group as one of the preeminent American philosophers of the twentieth century following his move to the University of Chicago in 1936, had, like Hahn, become a pacifist after being wounded in the war. He completed his Habilitation under Schlick with a thesis entitled "The Logical Structure of the World," which would become a classic of twentieth-century philosophy, applying Russell's and Whitehead's system of symbolic logic from *Principia Mathematica* to show how empirical facts can be built up into a rigorous image of reality. Logic was not just for logicians and mathematicians, but for "the whole of philosophy." Carnap insisted that by analyzing language as a set of rules of logical syntax which did not in themselves have meaning, the mistakes that allowed metaphysicians to be lulled by "verbal sedatives" into abstract concepts such as God or the unconscious would be vanquished. Carnap's first lecture to the Circle explaining the logical reconstruction of knowledge was, Feigl recalled, like listening to "an engineer describing the workings of a machine."[12]

Neurath, with his grand ideas for utopian reform, sarcastic invectives, and bohemian lifestyle (he lived in a fetid apartment, in a run-down building in a working-class district of Vienna), epitomized both the Circle's unconventionality and its undisguised contempt for philosophical traditions. He asked a philosophy student who came to see him what exactly he was studying.

"Philosophy, pure philosophy," the young man replied.

"How can you do something filthy like that?" Neurath shot back. "You might as well study theology!"[13]

Herbert Feigl conceded many years later that they had all been a bit too brash and cocksure, making needless enemies with their arrogant disdain for everyone who did not see the light as they did. "We thought we had found a philosophy to end all philosophies," he admitted.[14]

But "nothing was more odious" to the members of the Circle, said Menger, "than a hazy expression of presumed truths."[15] During the Circle's discussions, Neurath would constantly try to enforce discipline by banging his fist furiously on the table and interrupting if anyone dared utter one of the words on his long list of proscribed terms, such as "idea," "ideal," even "reality." He ultimately tried the patience even of the ever-patient Schlick, who would plead, "Dear Neurath, please permit us!" Neurath responded by making a small card with the letter "M" on it which he would silently hold up whenever the discussion strayed into the forbidden territory of metaphysics. After several weeks of this he announced, "I can abbreviate the procedure still further if I instead hold up a card with 'non-M' on it when you *don't* speak metaphysics."[16]

Hahn once wryly acknowledged, however, how far they had to go to convert the world to their view that all knowledge could be reduced to verifiable fact and pure logic. "If we were to open the window so that passersby could hear us," he remarked, "we would wind up either in jail or in the loony bin."[17]

UTTERING THE UNUTTERABLE

It was in fact Neurath's attempts to convert the world that pushed Gödel further away from the Schlick Circle in 1929. Taking advantage of Schlick's absence in America for a semester, Neurath dashed off a manifesto for what he conceived as a new phase of public outreach for the group, embracing an explicit social and political agenda. Neurath had been the chief instigator of a public offshoot of the Circle founded the year before, the Ernst Mach Society, which launched a series of popular books, pamphlets, and public lectures.[18] In his manifesto he claimed for Schlick's group the name "The Vienna Circle," declaring, "The scientific worldview serves life, and life embraces it," along with other slogans pledging to free the proletariat from the oppression of metaphysics, theology, and capitalism.

"All of us in the Circle were strongly interested in social and political progress," recalled Carnap. "Most of us, myself included, were socialists." Gödel was, too, by any usual definition of the word. He once told Carnap that he had read Lenin and Trotsky and was "for a planned society and socialism, and interested in the mechanism of influences in society, e.g., that of finance capital on politics."[19]

But, like Menger, Schlick, and even the ardent socialist Hahn, Gödel was strongly put off by Neurath's attempts to politicize their work.[20] Other enthusiasms of the Circle during this time confirmed for Gödel his sense that whatever intellectual sympathies he shared with its members began to diverge sharply at the place where mathematical logic ended and philosophy began. This had much to do with the Circle's rapturous admiration for the philosophical ideas of Ludwig Wittgenstein, particularly his thoughts on the role of language—which had seemed to

An intense Ludwig Wittgenstein

be exactly what Schlick, Carnap, Feigl, Hahn, and the others had been looking for when they first came across the cryptic aphorisms, as Feigl called them, of Wittgenstein's now-famous but then obscure *Tractatus Logico-Philosophicus*.[21]

They were among the first but definitely were not the last to fall under the spell of Wittgenstein's odd brilliance. The son of one of the wealthiest men of the Austrian monarchy, an iron and steel baron whose empire rivaled those of Krupp's and Andrew Carnegie's, he had grown up in Vienna in an opulent town *Palais* with marble stairways, a statuary-lined entrance hall adorned with frescoes depicting scenes from *A Midsummer Night's Dream*, and a street frontage stretching fifty yards along Alleegasse (now Argentinierstraße), with a great arched entryway at each end. Johannes Brahms and Richard Strauss were regular guests at musical evenings. Educated mostly by private tutors—except for an awkward and unhappy three years he spent at the Realschule in Linz, where one of his schoolmates for a year was the even more awkward and unhappy fifteen-year-old Adolf Hitler—Ludwig Wittgenstein had, like all his siblings, acquired self-reliance, social arrogance, and a difficulty with intimate relationships that would dog him throughout his life.[22]

Intending to become an engineer, he had traveled to Manchester, England, in 1908, at age nineteen, to work at an aeronautical research station where he hoped to design and build his own airplane. It was there that his growing interest in mathematics led him to the discovery of Russell's *Principia Mathematica*. With the passion of a convert, he set out for Cambridge, impetuously introduced himself to Russell, and announced his intention to make philosophy his life's work—if he could.

Russell at first was mostly annoyed by his young acolyte's wild enthusiasms. A few weeks after their first meeting he wrote to his lover Ottoline Morrell: "My German engineer very argumentative & tiresome." A day later: "My German engineer, I think, is a fool. He thinks nothing empirical is knowable—I asked him to admit that there was not a rhinoceros in the room, but he wouldn't." He took to haunting Russell in his rooms, for a while appearing every evening at midnight, and pacing back and forth "like a wild beast for three hours in agitated silence," Russell reported.

"Are you thinking about logic or about your sins?" Russell once asked.

"Both," Wittgenstein replied, and continued his silent pacing.[23]

At the end of the first term he came to Russell and demanded, "Do you think I am an absolute idiot?" Russell (in his later, and slightly improved account of their exchange) replied, "Why do you want to know?"

"Because if I am I shall become an aeronaut, but if I am not I shall become a philosopher."

"My dear fellow," Russell said, "I don't know whether you are an absolute idiot or not, but if you will write me an essay during the vacation upon any philosophical topic that interests you, I will read it and tell you."

As soon as Russell read the first sentence he knew the answer. As he later described Wittgenstein, "He was perhaps the most perfect example I have ever known of genius as traditionally conceived, passionate, profound, intense, and dominating," with a kind of unworldly "purity" that could make him both astonishingly rude and morally unshakeable.[24] When Russell's fellow Cambridge mathematician G. E. Moore gently tried to explain to Wittgenstein that an essay on logic, which Wittgenstein had dictated aloud to Moore in early 1914 in a remote hut on a Norwegian fjord where he had gone to live the life of a hermit for two years, could not be accepted as qualification for a BA degree—it lacked a preface on sources as required by the university's regulations—Wittgenstein erupted in typical moral outrage.

Dear Moore, Your letter annoyed me. *When I wrote Logik I didn't consult the Regulations*, and therefore I think it would only be fair if you gave me my degree without consulting them so much either! . . . If I am not worth your making an exception for me *even in some* STUPID *details* then I may as well go to HELL directly; and if I *am* worth it and you don't do it then—by God—*you* might go there.[25]

In his philosophical writings Wittgenstein had already begun the inquiries into language and logic that would so electrify the Vienna Circle—but alienate Gödel and Menger—when they first discovered them a decade later. The very first sentence of his essay on logic spoke squarely to this point:

> Logical so-called propositions *show* the logical properties of language and therefore of the Universe, but *say* nothing.[26]

And in the final line of the *Tractatus*, he repeated the central idea he had announced in the preface: "Whereof what cannot speak, thereof one must be silent."

With the outbreak of war in August 1914 Wittgenstein enlisted in the Austrian Army, less out of patriotism than the hope that "the nearness of death will bring me the light of life," he wrote in his diary. Decorated several times for bravery, he was serving as an artillery officer on the Southern Front when he was taken prisoner a week before the armistice. In the ensuing months spent in an Italian prisoner of war camp, he occupied his time trying to arrange for the publication of his *Tractatus*, which he told Russell had solved *all* of the outstanding problems in philosophy: "This may sound arrogant but I can't help believing it."[27]

The *Tractatus* took the form of a series of oracular pronouncements arranged in a numbered hierarchical structure. Wittgenstein rarely bothered to include any justification or argument, setting forth each proposition with an air of magnificent and unquestionable authority "as if it were a Czar's ukase," as Russell wryly put it. Between 1924 and 1927 the Circle worked its way through Wittgenstein's magnum opus twice, beginning to end.[28]

Schlick excitedly reported to Einstein that discovering Wittgenstein's ideas had been "the most tremendous intellectual experience of my life":

> I consider it to be altogether the deepest and truest book of recent philosophy. Reading it is, however, extremely difficult. The author,

who has no intention of ever writing anything again, is an unconventional type of ravishing genius. . . . His basic view seems to me
in principle to overcome the entire foundational crisis of current
mathematics.[29]

Though Schlick made repeated attempts to meet his new idol, Wittgenstein kept putting him off. He was then teaching school in a tiny
Austrian village, having earlier given away his fortune on the theory
that wealth was incompatible with the life of a philosopher—and, as
he put it, so he could "croak with self-respect." In April 1926, Schlick
and a few students took the long train ride out to the village of Otter-
thal to call on him, only to find he had abandoned teaching altogether,
having made himself unwelcome in one place after another with his
unconventional teaching methods ("that totally insane fellow who
wanted to introduce advanced mathematics to our elementary school
children," one villager recalled many years later) and his furious impatience with his students' lack of comprehension, which culminated in
his delivering a blow to the head of one of his less bright eleven-year-
old pupils, knocking him unconscious. By the time Schlick arrived,
Wittgenstein had decamped to a monastery outside Vienna, where he
was working as a gardener. Again, the attempted meeting failed to
take place.[30]

Finally, in late 1926, Wittgenstein was back in Vienna, now fan-
cying himself a modernist architect and hard at work designing a spacious townhouse for his sister, Gretl Stonborough, who managed to
arrange the long-awaited meeting with Schlick. "Each of us thought
the other must be mad," Wittgenstein said afterward. But Schlick was
subsequently able to lure Wittgenstein to join a select few from the Circle for Monday evening gatherings at his home or at a coffeehouse with
the promise that he would not be expected to talk philosophy—which
Wittgenstein made clear he had given up forever.[31]

It was a public lecture that Hahn arranged in March 1928 that
brought about Wittgenstein's momentous return to philosophy, the
logic of language in particular. The speaker was L. E. J. Brouwer, Hilbert's bête noire. "After much resistance," Feigl said, he and Waismann

managed to coax Wittgenstein into attending. Gödel was present for the great occasion, too. Hahn excitedly walked down the aisle to introduce himself as Wittgenstein entered the hall; Wittgenstein, as Menger recalled, "thanked him with an abstract smile and eyes focused at infinity, and took a seat in the fifth row." But at a coffeehouse afterward, Feigl reported, "A great event took place. Suddenly and very volubly Wittgenstein began talking philosophy—at great length." The dam had broken.[32]

Brouwer's radical view that mathematics is entirely a human construct was one Wittgenstein also fully shared. Brouwer dismissed the idea that mathematics in any sense constitutes an "objective" truth. In this he was diametrically opposed to the Platonist conception of mathematics as a body of ideal truths, preexisting somewhere "out there" independent of the human mind and awaiting human discovery.

Gödel scholars have debated at length how far back the undeniable Platonism of his later mathematical ideas went, some suggesting he may not have been as consistent on the point as he later claimed, when he wrote in 1976 that he had been a confirmed Platonist ("a conceptual and mathematical realist") "since about 1925."[33]

But there were ample signs even in these early years how very much out of sympathy Gödel was with the predominant views of the Vienna Circle, particularly when it came to the prevailing enthusiasm for Wittgenstein's ideas. Gödel told Carnap in 1928, in one of their many coffeehouse conversations, that he did not see why abstract mathematical concepts like infinity had to be justified on grounds of their empirical application to physical reality: he believed they possessed a reality of their own.[34] And he plainly disagreed with the reverence for Wittgenstein's idea that mathematics, like language, was merely a tool, a set of rules or a syntax that had no inherent meaning in itself. He would spend much of the decade of the 1950s writing and rewriting a chapter he had promised challenging Carnap's views on the matter, entitled "Is Mathematics Syntax of Language?" He abandoned it in the end not only because he was never satisfied with his endless revisions, but more important because he thought "basically it is a foolish question and trying to refute it gives it credit it doesn't deserve."[35] Replying to an inquiry late in his life, Gödel wrote,

> I *don't* consider my work a "facet of the intellectual atmosphere of the early 20th century," but rather the opposite. It is true that my interest in the foundations of mathematics was aroused by the "Vienna Circle," but the philosophical consequences of my results, as well as the heuristic principles leading to them, are anything but positivistic or empiristic. . . . So my work points toward an entirely different world view. . . . Also the interest of the Vienna Circle in the foundations of mathematics began long before Wittgenstein's Tractatus.[36]

However elusive the concept of mathematics was, Gödel throughout his whole life would devotedly regard it as a search for truth, and more specifically a search for pre-existing truths that inhabited a reality separate from the human mind. It was a philosophical worldview that had much more in common with the nineteenth, or even the seventeenth century, than the twentieth, and could not have been more at odds with that of Wittgenstein and the Vienna Circle.

Even the more enthusiastic Wittgensteinians within the Circle, however, at times poked fun at Schlick's cultlike adoration of his idol. Gödel's friend Natkin wrote to him in July 1927, "As a consolation, I send you Schlick's essay, an example of how only about meaningless things can one speak meaningfully. I don't know if Feigl told you about Schlick's conversation with Wittgenstein, in which they entertained themselves for hours discussing the inexpressible."[37]

By 1929, Gödel had stopped coming regularly to the Circle's Thursday evening meetings. After one session in which Schlick, Hahn, Neurath, and Waismann had talked at great length about language, but neither Gödel nor Menger said a word, Menger remarked to his younger companion on the way home, "Today we have once again out-Wittegensteined these Wittgensteinians: we kept silent."

"The more I think about language," Gödel replied, "the more it amazes me that people ever understand each other."[38]

ENDEARING INNOCENCE

During the 1928–29 academic year, at the request of several of his students, Menger began his own seminar devoted more strictly to mathematical topics. Gödel found Menger's *Mathematisches Kolloquium* a much more comfortable setting to express his ideas, and opened up to his colleagues there in a way he never did at Schlick's Circle. Menger recalled:

> Towards the end of 1929 I also invited Gödel to the Colloquium and from then on he was a regular participant, not missing a single meeting when he was in Vienna and in good health. From the beginning he appeared to enjoy these gatherings and spoke even outside of them with members of the group, particularly with Georg Nöbeling, sometimes with Franz Alt and Olga Taussky, when she was in Vienna, and later on frequently with Abraham Wald and foreign visitors. He was a spirited participant in discussions on a large variety of topics. Orally, as well as in writing, he always expressed himself with the greatest precision and at the same time with the utmost brevity. In nonmathematical conversations he was very withdrawn.[39]

Olga Taussky also found him cautious to the point of withdrawal in ordinary social situations. She often invited him to the gatherings for tea at her family's apartment in Vienna that she held for members of her circle of mathematical friends from the university and occasional visiting scholars. While Gödel gave "the impression that he enjoyed lively people," Taussky said, he "was very silent" and "did not contribute himself to nonmathematical conversations." But on occasions when he was alone, she added, "you could talk to him about other things too and his clear mind made this a rare pleasure."[40]

Gödel revealed some of his underlying social anxieties in his 1937–38 diary, especially his dread of revealing his inner self and failings to the scrutiny of others:

At tea with Olga Taussky, Karl Menger, and foreign visitors

Why is it so embarrassing to me that other people know everything about me? Because it is the basis for a value judgement and each value judgement is embarrassing to me, either

1. It is too favorable (they expect something good of me, hold me in esteem) then it is embarrassing due to disappointment
2. They know the truth, then no esteem which is also embarrassing

He berated himself for his perceived shortcomings: lying in bed too late in the morning, overestimating himself and underestimating others, "laughing, talking in a high voice, making faces," indecisiveness, "nerves and insecurity in company." He concluded: "These are numerous 'demerits' of the person which would encourage no one to want to be like me (rather the opposite) and which secondly make it seem impossible for me to accomplish anything for the good of others." But then he added: "Oddly enough the embarrassment is diminished by clearly articulating all of one's bad qualities."[41]

But with the friends whom he did feel comfortable with he was an entirely different person, warm, gently humorous, loyal, and conscientious, always willing to assist with their mathematical ideas while bashfully modest about his own work. As Menger recalled:

> Gödel obviously continued to enjoy the meetings of the Colloquium, spoke with the other participants and was generous with his advice in logical and mathematical questions. He always grasped problematic points quickly and his replies often opened new perspectives for the enquirer. He expressed all his insights as though they were matters of course, but often with a certain shyness and a charm that awoke warm and personal feelings for him in many a listener.[42]

That "endearing innocence," as Karl Sigmund aptly describes it, drew many to him. Gödel's last years in Princeton as a reclusive loner have cast a long shadow over the received portrait of his personality. But that is not the Kurt Gödel his friends knew throughout his life, right up until those final tragic years of isolation and paranoid delusion. Gödel, said his close friend Feigl, was "very unassuming" even though it was apparent to all that "his was clearly the mind of a genius of the very first order." Feigl, whose Jewish but ardently atheistic father had been a skilled weaver and subsequently a leading figure in the Bohemian textile industry, had grown up in a decidedly secular and cultured home, and had gone on to study physics, philosophy, and mathematics at the University of Munich. With their other close friend from the Circle, Marcel Natkin, a philosophical prodigy from Lodz, they "met frequently for walks through the parks of Vienna, and of course in cafés had endless discussions about logical, mathematical, and epistemological, and philosophy-of-science issues—sometimes deep into the hours of the night," Feigl recalled. The three would meet again for an emotional reunion in New York in 1957, a quarter of a century after they had gone their separate ways.[43]

"His friendliness & his quiet humor are so appealing," Gödel's later friend Oskar Morgenstern recorded in his diary. Morgenstern was always touched by Gödel's solicitous concern for his family, and the hours he would generously spend talking to and encouraging Morgenstern's son Carl in his interest in mathematics as a high school and college student, along with the "presence of greatness of mind" that always came through that unassuming exterior: "Talking to him (even before leaving for Vienna) is always an experience. The amazing cheerful sense, quite apart from what is said. Hardly anyone believes that because he always looks so serious."[44]

Rudy Rucker, a mathematician and science fiction writer who made the "pilgrimage" to call on Gödel in Princeton several times, described his "conversation and laughter":

> His voice had a high, singsong quality. He frequently raised his voice toward the ends of his sentences, giving his utterances a quality of questioning incredulity. Often he would let his voice trail off into an amused hum. And, above all, there were his bursts of complexly rhythmic laughter.[45]

The humor that came through easily to intimate friends, and in his letters to his family, was indeed not always so easy for others to pick up. Gerald Sacks, a Harvard logician who was a visitor at the Institute for Advanced Study in 1961–62 and again in 1974–75, recalled catching a fleeting and amusing glimpse of it when he tried challenging Gödel about the ideas of his philosophical idol, Gottfried Leibniz.

> I asked him, how about monads—because I knew for him the great philosopher was Leibniz. I said, I find this theory very strange that the universe is composed of monads, which have no information about each other—it seems to contradict common sense, because we see this great harmony. Leibniz's answer was that God imposed this harmony. So, I asked him what he thought. He said, "Leibniz was wrong. It's obvious, everything that happens affects everything else. This is part of science." And he said, "Furthermore, Leibniz was wrong about everything."

Then he paused and said, "But it's just as hard to be *wrong* about everything as to be *right* about everything."

He didn't crack a smile, but I thought, this was a very amusing remark. And I thought, Wow, he cracked a joke. Or did he?

Even at that late date, Gödel had a "zest for life" and enthusiasm for ideas, Sacks saw. "He was full of wonder and excitement over developments in math, logic, philosophy." It was, Sacks said, like talking to "a very bright eleven-year-old."[46]

That innocence engendered a feeling of protectiveness among more than one of those who came to know and care for him. Notably, Karl Menger and later John von Neumann and Oskar Morgenstern, all of whom were just a few years older than Gödel, looked after him with an almost fatherly solicitousness and affection. Menger and von Neumann could easily have viewed this *Wunderkind* as a professional rival, but instead went to often extraordinary lengths to save him from his own naiveté in the world, ensuring that he had the opportunity and financial support to carry on his work, which they both did much to make known throughout the world of mathematics. Morgenstern patiently saw him through his many crises and used his influence repeatedly to have him accorded the recognition and honors he deserved.

Gödel's boyish charm was not lost on girls, either. For all of his shyness and social fears, he could, as Olga Taussky once observed, be a bit ostentatious about his success in attracting the opposite sex. Once Taussky was working in the small seminar room at the Mathematical Institute when a very pretty, beautifully dressed girl came in and sat down, followed a few minutes later by Gödel, who ceremoniously departed with her. "It seemed a clear show off on the part of Kurt," Taussky said.[47]

That girl soon gave up Gödel, whose "prima donna" habit of sleeping late in the morning she had unsuccessfully tried to reform. But another, older woman did not give up so easily. In 1928 Gödel and his brother moved to a third-floor apartment at Lange Gasse 72. Living almost directly across the street, at Lange Gasse 67, was a twenty-nine-year-old woman named Adele Nimbursky, whose father was a failed

Top: With Adele at an outdoor café in Vienna
Above: Adele on stage, age nineteen

artist, though apparently more successful as a photographer. Separated from her husband, she had recently moved back with her parents after a brief and unhappy experience with married life. She had once worked as a dancer in a nightclub, Der Nachtfalter—"The Night Moth"—and was now trying to make a living offering massage and foot care services. Neither her past nor her present occupation was quite as dis-reputable as it sounded: while there was undoubtedly no shortage of Viennese dancers and masseuses in the 1920s and 1930s who were prostitutes, there were also those who were most definitely not, and Adele listed her services in the city directory at the address of her very strict and conservative Catholic father.[48]

Over the following ten years Gödel would find himself, at times against his inclination, and definitely against the wishes of his family, increasingly under the thrall of this uneducated but determined woman nearly seven years his senior.

SHAKY FOUNDATIONS

In deciding to take on the fourth of the challenges Hilbert had put forth at the Congress of Mathematicians in 1928, Gödel placed himself at the very center of the storm over mathematical foundations, which had broken with a deeply unnerving discovery Bertrand Russell had made at the turn of the century while working on *Principia Mathematica*. Russell's idea had been to establish the soundness of mathematics by show-ing how it could all be reduced to principles of logic so self-evident as to be beyond doubt. Defining even the simplest operations of arithmetic in terms of what Russell called such "primitive" notions, however, was far from an obvious task. Even the notion of what a *number* is raised immediate problems. The laboriousness of the methodology and nota-tion was all too evident in the (often remarked) fact that it took more than seven hundred pages to reach the conclusion, "1 + 1 = 2," a result which Russell and Whitehead described as "occasionally useful."[49]

One way to reduce the abstract notion of numbers to logical con-creteness was through set theory. A set, or "class" as Russell called it, is simply a group of things. The elements that make up a set can be either explicitly listed or defined by a shared property: the set of black cats; the set of primary colors; the set of the numbers 1, 7, and 23; the set of all odd numbers; the set of failed presidential candidates. Draw-ing on the earlier work of the German mathematician Gottlob Frege, Russell proposed that the way to think of the number 2 is simply as the definition of all sets that contain a pair of objects, and he proceeded to construct the rest of the numbers from there.

That innocent beginning swiftly led to dark waters. Sets can also be made up of other sets: for example, a set whose elements are defined as a married couple might be {Napoleon, Josephine}; the set of all such sets would be { {Napoleon, Josephine}, {Elvis, Priscilla}, {Bill, Hillary}, {Bob, Carol}, {Ted, Alice}, etc. }. Some sets, Russell observed, could even be members of themselves. For example, the set of all sets containing more than two elements would certainly be such a set, as it itself contains more than two such sets among its own members. By way of analogy, a Wikipedia article with the title "List of Wikipedia articles whose title begins with the letter L" would itself belong on that list. The unanticipated calamity came when Russell asked the seemingly innocent question, What about the set of all sets that are *not* members of themselves? Is *it* a member of itself, or not? If it is not, then by definition it should be included. But if it is included, that means it *is* a member of itself, so should not.

"Russell's Paradox," as it came to be known, echoed paradoxes that had been around since antiquity. The prototype is the Liar's Paradox, attributed to Epimenides the Cretan, who asserted, "All Cretans are liars." Russell noted that this was akin to the conundrum posed by a piece of paper on which the sentence "The statement on the other side of this paper is false" is written on one side, and the sentence "The statement on the other side of this paper is true" on the other.

"It seemed unworthy of a grown man to spend his time on such trivialities," Russell later recalled, and "at first I supposed that I should be able to overcome the contradiction quite easily, and that probably there was some trivial error in the reasoning." The more he thought about it, the more he realized it was a flaw in his whole project too deep to be ignored. He almost destroyed Frege when he pointed it out to him just as Frege was about to publish his entire life's work on mathematical foundations. Frege published a gracious addendum that Russell later called almost "superhuman," acknowledging that "his fundamental assumption was in error," and thanking Russell for his discovery.[50]

But the problem remained. As Frege ruefully observed, citing the

Latin proverb that means "It is comfort for the wretched to have company in his misery,"

> *Solatium miseris, socios habuisse malorum.* I too have this solace, if solace it is; for everyone who in his proofs has made use of extensions of concepts, classes, sets, is in the same position. It is not just a matter of my particular method of laying the foundations, but of whether a logical foundation for arithmetic is possible at all.[51]

The unsettling possibility that mathematics had no logical foundation because logic itself was flawed was summarized by Robert Musil in an essay he wrote in 1913, "Der mathematische Mensch." Its lyricism could not soften how unnerving the situation was:

> And suddenly, just as everything had been brought into the prettiest existence, the mathematicians—those who contemplate to the very innermost—came to the conclusion that there was something at the foundation of it all they simply could not put right; indeed, they looked down and found that the entire structure was floating in midair.

Musil added, "The mathematician carries this intellectual scandal in an exemplary manner, that is, with confidence and pride in the devilish audacity of his own mind."[52]

Russell, arguably with a hint of such audacity, resolved his paradox finally by outlawing it: his "theory of types" simply decreed that no class is permitted to refer to itself. There was a certain logic to his edict, to be sure: like a book listing all of the books in a library, *that* book did not yet exist when it was being prepared, so it could not be expected to contain itself. But many found Russell's solution ultimately unsatisfying, including Gödel, who called it "too drastic" a measure. "He cured the disease," agreed the prominent German mathematician Hermann Weyl, "but . . . also imperiled the very life of the patient."[53]

But there were other problems with the entire exercise that Russell himself recognized, and not long after finishing his herculean task he admitted feeling a deep sense of failure. The massive manuscript, with its complex notation which could only be written out laboriously by hand, had to be carted in a four-wheeler cab to the offices of Cambridge University Press when it was finally done. The press was so pessimistic about the sales prospects that it demanded £600 to cover the anticipated loss; even with a grant from the Royal Society and other funding, Russell and his co-author Whitehead ending up earning "minus £50 apiece for ten years' work," Russell said. "This beats the record of *Paradise Lost.*" A number of years later he told a friend, with his usual self-mocking humor, "I used to know of only six people who had read the later parts of the book."[54]

One of those was Kurt Gödel, who in the summer of 1928 spent 385 Czech crowns (about $150 in the currency of the early twenty-first century) to purchase a copy of the first volume, dutifully spending the rest of the summer at his parents' home in Brno reading it. He, too, was disappointed. "I was less enthusiastic than I had expected from its reputation," he wrote Feigl at the end of the summer.[55]

Aside from all else, Russell's program of grounding mathematics in logic—the school of foundational thought known as *logicism*—was, in the larger sense, at best a partial success, since at least two of the axioms of set theory it depended upon were far from self-evident logical propositions at all. Some, to be sure, fit the bill, such as the assertion that a set can contain no elements. But the Axiom of Infinity, which postulates the existence of sets with infinite numbers of elements, and the Axiom of Choice, which asserts that a function always exists that will specify exactly one element from each set within an infinite set of sets, both relied on exactly the kind of leap into the intangible that was now coming under withering attack from the flanks of Hilbert's optimistic program for securing the foundations of mathematics, and from none other than Hilbert's old nemesis Brouwer.[56]

GOD NEVER DOES MATH

Hilbert's approach, which he termed *formalism*, sought to slay Russell's paradoxes by a different means altogether. If the methods of mathematics itself could be used to establish that the process of proof operates without throwing up any contradictions, then, Hilbert argued, the entire mechanism could safely be declared sound. Hilbert called his method "proof theory"; *metamathematics* was the term later used to describe this mathematical examination of mathematics itself. "My investigations on the refounding of mathematics aim at nothing less than to definitively eliminate from the world the general doubts about the certainty of mathematical reasoning," he said. The present situation, however, was simply "intolerable." Where else were reliability and truth to be found, he asked in exasperation, if "even mathematical thinking fails?"[57]

Wittgenstein for his part made light of all such hand-wringing. As he later expressed it, "What does mathematics need a foundation for? . . . The *mathematical* problems of what is called foundations are no more the foundations of mathematics than the painted rock is the support of the painted tower." As far as he was concerned, all of the work on foundations was just irrelevant prettification.[58]

Expanding on Wittgenstein's argument that logic is nothing but a series of tautologies—which merely restate one universally valid proposition in terms of another, without adding any new knowledge about the world—Gödel's dissertation adviser Hans Hahn now suggested that the same could be said of mathematics in its entirety. "It seems hardly credible at first sight that the whole of mathematics with its hard-earned theorems and its frequently surprising results could be dissolved into nothing but tautologies," Hahn acknowledged. "But this argument overlooks just one minor detail: it overlooks the fact that we are not omniscient." An omniscient being could instantly perceive that "24 × 31" and "744" were merely two different ways of expressing the same thing, an equivalency that humans had to laboriously work out: "An omniscient being needs no logic and no mathematics." Plato had said, "God is always doing geometry." Hahn countered, "God *never* does mathematics."[59]

All of this was cold comfort to Hilbert, particularly in the face of the attacks on formalism now emanating from Brouwer's camp. While

Hilbert sought to convince the world that the buildings mathematicians had been erecting for years were perfectly safe and trustworthy, Brouwer's solution to the foundational crisis was to tear the whole rotten edifice down and start over again from the ground up, this time with a much stricter building code. The end result would be a much smaller but, he insisted, far more resilient structure.

That was the topic of his talk in Vienna that Wittgenstein had been cajoled into attending. Departing both from Russell's logicism and Hilbert's formalism, Brouwer's *intuitionism*, an approach more generally known as *constructivism*, rejected a large body of modern mathematics altogether. Brouwer had long been the enfant terrible of mathematical logic. In 1908 he had written a paper entitled "The Unreliability of Logical Principles," which rejected two thousand years of logical reasoning, back to Aristotle. When he wasn't picking fights with his fellow mathematicians, he lived and worked in a small garden shed furnished with nothing but desk, bed, and a piano in an artist's colony outside of Amsterdam.[60]

The constructivists insisted that since all mathematical objects and concepts, from integers to sets to geometric shapes, are human inventions, they need to be substantiated with explicit recipes for their construction. The constructivists particularly objected to what they regarded as the profligate and reckless misuse of the Law of the Excluded Middle in mathematical proofs. A standard procedure of proof is to assume the *opposite* of what is to be demonstrated, and show that this leads to a contradiction. Such "non-constructive" proofs abound in mathematics, and are often of elegant simplicity. To prove that in a room of 367 people at least two people must share the same birthday, it is not necessary to work through every possible combination of birthdays, but simply show that if that were *not* the case an impossibility results, since if all 367 had different birthdays, that would mean there are more possible birthdays than there are days in the year. Euclid, in 300 B.C., used the same method to prove that there is no largest prime number. A prime number is divisible by only 1 and itself; a number that is not prime can always be expressed as the product of two or more primes. So 9 is not prime because it can be expressed as 3×3; 50 is not prime because it can be expressed as $2 \times 5 \times 5$. Euclid's argument ran like this: Assume there *is* a largest prime number, p. You can then create a larger number by multiplying together every prime number up to and including p, and adding 1. But that number, when divided by *any* of those primes, will leave a remainder of 1—so it, too, is prime (or, possibly is divisible by a prime larger than p, likewise demonstrating that p is not the largest prime).

When finite sets of numbers are involved, the constructivists were willing to admit such reasoning based on the Law of the Excluded Middle, which states that either A or not-A can be true, but not both. Thus proving the impossibility of not-A is sufficient to establish A. But Brouwer completely rejected the idea of applying the law to infinite sets. In that situation, consistency is not enough to establish existence, he insisted. His favorite example, one that Wittgenstein later often used, was to ask how one could possibly prove or disprove the proposition that somewhere in the infinite decimal expansion of pi the sequence 7777 occurs. Not finding it could always be answered with the objection that one has just not looked far enough.[61]

Russell had made an analogous point with his whimsical example, "The present King of France is bald." The sentence would appear to be false, since there is no King of France at present, but that would not warrant applying the Law of the Excluded Middle to conclude, "The present King of France is *not* bald."[62] Brouwer's point was that when dealing with infinite sets, use of the law similarly assumes the existence of things we have no right to assume. As Gödel himself would later note, non-constructive proofs which establish the existence of mathematical properties through such indirect means often have "strange results . . . e.g., that we can often prove the existence of an integer with a given property without anyone's being able actually to name such an integer or even to describe a procedure by the application of which we could obtain such an integer."[63]

Hilbert responded to Brouwer's attacks with his usual call to arms.

"Taking away the Law of the Excluded Middle from the mathematician," he cried, "would be like forbidding the astronomer his telescope or the boxer the use of his fists!" And, referring to the work of Georg Cantor, whose studies of infinite sets in the late nineteenth century were regarded as a milestone in set theory and mathematical foundations, Hilbert vowed, "No one shall drive us out of the Paradise that Cantor has created for us!"[64] When Brouwer gave a talk at Göttingen presenting his iconoclastic ideas, Hilbert stood up at the end and drily responded, "With your methods, most of the results of modern mathematics would have to be abandoned, and to me the important thing is not to get fewer results but to get more results." More acerbically, he dismissed Brouwer as the leader not of the revolution he claimed, but of a pathetic little *Putsch*—"now, with the State armed and strengthened . . . doomed from the start!"[65]

Brouwer shortly afterward riposted that "formalism has received nothing but benefactions from intuitionism. . . . The formalist school should therefore accord some recognition to intuitionism, instead of polemicizing against it in sneering tones." He pointedly added, "In the framework of formalism, *nothing* of mathematics proper has been secured up to now." In his 1928 lecture in Vienna he had indeed offered an important benefaction, in the form of a deep insight that went to the heart of Hilbert's methodology for establishing the soundness of the formal mechanism of mathematics and which would carry enormous implications for Gödel's subsequent earthshaking results. As Brouwer noted, there is a fundamental distinction between demonstrating that a system is *correct* ("richtig"), that no false proposition can ever be derived from its axioms, and demonstrating that it is *consistent* ("nichtkontradiktorisch"), that it is impossible to derive from the axioms both a proposition and its negation, both *A* and not-*A*. The former is a *semantic* requirement, the latter only *syntactical*. The corresponding pair of concepts belonging to Hilbert's challenge to demonstrate the completeness of a formal system are *completeness* and *decidability*. A system is strictly speaking *complete* if every true statement can be derived from its axioms; it is *decidable* (or "syntactically complete," or "negation-complete") if for any proposition *A*, either *A* or not-*A* can always be derived.*[66]

By Brouwer's definition, Hilbert's formalist program was purely syntactical, a profoundly clarifying insight. Hilbert was in effect proposing that there is no need to run a quality-control check on every single product emerging from the factory; the aim, as Gödel later explained it, is "to guarantee the reliability of the mathematical apparatus." For that, it is sufficient to show that the machinery—the axioms and rules of inference that operate within a mathematical system—is working as designed, the gears turning without jamming.[67] For the purposes of proof theory there is thus no need to establish *which* of the two assertions *A* or not-*A* is actually true in any particular case; it is sufficient merely to establish that the mechanism is always guaranteed to prove one or the other—and never both.

The key point, as Gödel subsequently explained it, is that testing the machinery requires examining only the system's syntax, not its meaning. "We don't have to worry about the meaning of the symbols of our system because the rules of inference never refer to meaning," he noted. For example, if we prove *A*, and also that *A* implies *B*, we are entitled to conclude *B*, without having to even bother about what *A* and *B* actually stand for. The rules are "purely formal," Gödel noted: "They could be applied by someone who knew nothing about the meaning of the symbols. One could even easily devise a machine which would give you as many correct consequences of the axioms as you like."[68]

Gödel's mentioning a "machine" was a suggestive anticipation of the profound connections his work would in fact have to the theories of Alan Turing and John von Neumann, laying the basis of the digital computer a decade later. It was just one of many far-reaching ripples from the series of rocks he was about to throw into the waters of mathematical logic.

---

* The term "decidable" also has another sense when referring to the so-called *Entscheidungsproblem*, which asks whether a computer program or other algorithm can, in a finite number of steps, determine whether a proposition is true or false. Alan Turing demonstrated in 1936 that no such algorithm can be constructed.

· 5 ·

Undecidable Truths

ONE CHEER FOR COMPLETENESS

Gödel apparently never even showed his dissertation to Hans Hahn, his adviser, before submitting it around February 1929, barely half a year after Hilbert issued his challenge in Bologna.[1] In proving the completeness of first-order logic, he had at one stroke done more to advance the formalist program than any member of the "younger generation of mathematicians" whose cooperation Hilbert sought to enlist.

Gödel's thesis, "On the Completeness of the Calculus of Logic," was formally approved by Professors Hahn and Furtwängler on July 13, 1929, and the following February, at age twenty-three, he was granted his doctorate. The proof itself is remarkably brief. When it was published later that year in *Monatshefte für Mathematik und Physik*, it occupied just twelve pages.[2]

First-order logic builds upon the basic logic of propositions by supplementing the logical connectors *and, or, not, if . . . then,* and the like with symbols that permit propositions about "some" or "all" objects. It does this with two so-called quantifiers, first introduced by Frege. If x is any member of the set of cats, and the function $F(x)$ stands for the proposition "x is black," then the statement "some cats are black" can be written using the "existential quantifier" $\exists$ as

$$(\exists\, x)\ F(x)$$

which literally asserts, "There exists some value of x for which the proposition $F(x)$ holds." The proposition "all cats are black" can likewise be expressed by employing the "universal quantifier" $\forall$ to form the sentence, "For all values of x, $F(x)$ holds." In first-order logic, "some" or "all" statements can be made about objects (all cats, some numbers, some sets), but not about properties or sets of those same objects, which rules out such propositions as "there is some color that two cats share" or "some sets of cats contain a black cat." Despite that limitation, it is a powerful system, the basis of much of the mathematical foundations Hilbert sought to secure.

Gödel in his proof was able to demonstrate not only that the axioms of the system are sufficient to guarantee that all universally valid logical propositions can be derived from them, but also that all the axioms are independent of one another: none can be derived from any of the others, so none can be dropped.

Although Gödel's lasting fame was to come from his Incompleteness Theorem that would burst upon the mathematical world just a few short months later, his comparatively less well-known Completeness Theorem is still viewed by some logicians as the more important result. "The Completeness Theorem is very powerful, and year after year there are surprises drawn from it," Gerald Sacks observed in 2007. "It allows you to create very, very strong mathematical structures from very weak hypotheses. And it is in fact a universal method."[3]

Given Hahn's not exactly overwhelming statement approving Gödel's thesis, he may have failed to appreciate its full import at the time: "The paper is a valuable contribution to logical calculus, fulfils in all parts the requirement for a doctoral thesis, and its essential parts deserve to be published," his adviser wrote.

As Gödel was in the midst of completing his thesis, his father suddenly died, on February 23, 1929. He had been suffering from a painfully enlarged prostate; surgery, which in that pre-antibiotic era and with the primitive techniques employed at the time had a 20 percent mortality rate, led to a fatal infection. He was just a few days shy of his fifty-fifth birthday. "With that collapsed for all of us, but particularly my mother, an entire world," said Rudi Gödel.[4] Kurt's friend Herbert Feigl wrote him when he heard the news a few weeks later:

Dearest Gödel!

I only today found out about the terribly tragic event that you have experienced. I need not tell you that I sincerely share your pain. Poor lad, you must have gone through a difficult time.

Please tell me if you come to Vienna again, I would of course like to speak to you. If I can be helpful to you in any way, share it with me.[5]

Although Gödel must have returned home for the funeral, given Feigl's letter addressed to him in Brünn, no letters from him from the time describing the event or his own feelings survive. In the diary in which he recorded some of his innermost thoughts seven years later, he upbraided himself for the emotionlessness he had felt toward his father, among a list of other perceived failings. But he scarcely mentioned his father again, to his family or anyone else, and did not even keep a picture of him. When, three decades later, his mother sent him an old photograph she had found, he answered, "Papa looks astonishingly young in that picture. That is not how I remember him at all. His hair in the last years was already heavily greying, whereas in this picture it is still quite black. . . . At any rate I am happy to have at least *one* picture of Papa here."[6]

Three days after his father's death Gödel cut his final ties to the place of his birth, formally renouncing his Czech citizenship to begin the process of becoming a citizen of Austria.[7] Gödel's mother was left with the villa in Brno, a modest pension from the Redlich company, and a sizeable inheritance from her husband, which made her and both of her sons well off, at least for the moment. That November she moved to Vienna to join them, and all three rented a large apartment together on Josefstädter Straße, where they would live, together with their father's foster mother, Aunt Anna, for the next eight years. It was just five hundred feet from the famous Josefstadt Theater and a quarter of a mile from the Café Reichsrat, with a ritzy cinema on the ground floor.

With her pension of about 300 schillings, or $50, a month, plus other investments, his mother's monthly income came to about $300, the equivalent of roughly $5,000 in the currency of the early twenty-first century—a sum which just met her monthly living expenses. Kurt's monthly budget was about the same. A note he made in the mid-1930s indicated that the family held bank account balances totaling some 200,000 schillings, or about $30,000, plus several thousand dollars' worth more in stocks and other investments. All told, it was the equivalent of about half a million 2020 dollars—a comfortable though hardly inexhaustible nest egg.[8]

But for the immediate future, the money gave him the freedom not to worry about making a living while he tried to make a name for himself, with the hope that by doing so he might eventually work his way into one of the very few and hard-to-come-by academic positions in the field in Austria. That, as he later said, was all he ever wanted: to obtain the financial security to be able to continue doing the work he loved.

A SUMMER'S NIGHT AT THE CAFÉ REICHSRAT

In September 1930 a huge scientific gathering was slated to take place in Königsberg, the port city of Hilbert's birth on the eastern Baltic. Three major German professional societies were holding their annual meetings simultaneously: the German Mathematical Society, the German Physical Society, and the Society of German Natural Scientists and Physicians. During the last such gathering of German scientific societies, in Prague the year before, the Vienna Circle had organized a "satellite" meeting as part of its new effort at international outreach and public promotion. Hahn, Waismann, Neurath, Feigl, Carnap, Philipp Frank, Richard von Mises, and the other denizens of the Circle had all presented papers. Encouraged by that success, they were now planning an encore performance on September 5–7 in Königsberg to bring the gospel of logical empiricism to the scientists assembled there.[9]

The highlight of the planned Second Conference on the Epistemology of the Exact Sciences was an opening session in which representatives of the three competing schools of mathematical foundations would each make their case. Russell and Brouwer were not planning to come to Königsberg, and Hilbert was fully involved with the major societies' conferences across town, but each of the chosen speakers was an able champion for their absent leaders: Carnap was to speak for the logicists, John von Neumann for the formalists, and Brouwer's most capable student, Arend Heyting, for the intuitionists. Waismann would then follow on behalf of Wittgenstein's ideas regarding the tautological nature of mathematics.[10]

As usual when Wittgenstein was involved with anything, even tangentially, he immediately caused trouble. He finally agreed that Waismann could speak about his views, but only if he made a statement at the outset of his presentation to the effect that "Ludwig Wittgenstein had declined any responsibility whatsoever for any view that Waismann might wish to attribute to him."[11]

Gödel was scheduled to give a presentation on the second day on the subject of his recently completed dissertation, "Completeness of the Logical Calculus." A few days before their departure, Gödel and a few of the others going to the conference met at the Café Reichsrat on a pleasant summer evening, Tuesday, August 26, to discuss their travel plans. Carnap, Feigl, and Waismann were there; they arranged to take the train to Swinemünde then go by steamer to Königsberg.

Six months earlier Gödel had already dropped a hint to Carnap that he was forging ahead to the next, much more difficult phase of Hilbert's program: extending the proof of consistency and completeness to a mathematical system like that of Russell and Whitehead's *Principia Mathematica*, which comprehends all of arithmetic and number theory. He had begun apparently by first trying to establish the completeness of higher-order logical systems.[12] He had also dropped a hint that there was a possible monkey wrench in the whole works. Carnap recorded in his diary two days before Christmas 1929:

> Café Arkaden 23.12.29. 5.45–8.30. Conversation
> with Gödel. On the inexhaustibility of mathematics.
> He was stimulated to this by Brouwer's Vienna lec-
> tures. Mathematics is not completely formalizable. He
> appears to be right.[13]

On that August evening Gödel revealed for the first time to another human being that he had solved the problem. Carnap's brief diary nota-tion is the only record of that moment in mathematical history:

> Café Reichsrat Tue. 26 6–8:30. Gödel's discovery:
> incompleteness of the system of PM [Principia Mathe-
> matica]. Problems with the proof of consistency.[14]

Three days later they met again at the Café Reichsrat and Gödel offered further details. But it was not until the very final minutes of his own presentation at Königsberg that Gödel made the first pub-lic allusion—"announcement" would be much too strong a word for his brief passing reference—to his great discovery. After thoroughly describing his Completeness Theorem, he concluded by matter-of-factly stating:

> If the completeness theorem could also be proved for higher parts
> of logic (the extended functional calculus), then it could be shown
> with complete generality that decidability follows . . . so the solv-
> ability of every problem of arithmetic and analysis expressible in
> *Principia Mathematica* would follow.
>
> Such an extension of the completeness theorem is, however, impos-
> sible, as I have recently proved; i.e., there are mathematical problems
> that can be expressed in *Principia Mathematica*, which cannot be
> solved by the logical means of *Principia Mathematica*. However it
> would take us too far afield to go into these things more closely.[15]

In an early draft of his lecture, Gödel had planned to conclude with a slightly different line: "These things, however, are still too little worked through to go into more closely," suggesting that his result had come together in the intervening time, just a few weeks before the conference. The day following his talk, in the conference's concluding discussion session on mathematical foundations, he provided a few more details. "One can, assuming the consistency of classical mathematics, even give examples of propositions (and in fact of the kind of Goldbach or Fermat) that really are contentually true but are unprovable in the formal system of mathematics," he said.[16]

"Goldbach" and "Fermat" were allusions to two famous unsolved propositions in mathematics that although they appeared to be true, had for centuries stubbornly defied all attempts at formal proof. In 1742, Christian Goldbach, an amateur mathematician who also studied law and medicine, wrote a letter to the great mathematician Leonhard Euler in which he proposed that every even number greater than 2 is the sum of two primes. Although no counterexample has ever been found (and computers have tested numbers up to seventeen digits long), no proof has been found either. The previous century, the French mathematician Pierre de Fermat, in a sly marginal note written in his copy of Diophantus's *Arithmetica*, claimed to have a proof for the theorem that no three positive integers a, b, and c can satisfy the equation $a^n + b^n = c^n$ for any integer n greater than 2. (Fermat in his note said that his proof was too long to fit in the margin. The theorem was finally proved in 1994, 357 years after it was first posed, in a proof that ran to 129 pages.)

As Gödel pointed out, the existence of undecidable propositions was not just a matter of incompleteness but a threat to the integrity of the whole works. If a statement F were true but not provable, then the statement not-F would not cause any contradictions to arise if it were added as an axiom to the system, as there was nothing in the system to disprove it. In such an unsettling circumstance, "One obtains a consistent system in which a contentually false proposition is provable."[17]

Hahn asked Gödel to provide a short account of this discovery for inclusion in the published account of the discussion, which appeared in the next issue of the Vienna Circle's journal *Erkenntnis*. His complete proof was published in *Monatshefte für Mathematik und Physik* the following year.[18]

As details of his accomplishment began to spread through the mathematical world, it was soon apparent that the proof itself was a feat of mathematical wizardry so extraordinary that its conception and execution were as awe-striking as the result. There are works even of genius where one can picture the intellectual scaffolding that built it up step by step; there are others, the hallmark of only a few of the very greatest works of art, that carry an eerie sense of having been born whole, with all of their parts falling in place just where they are needed, in a way that seems to defy human anticipation. Like the countersubjects of a Bach fugue that work perfectly together at the moment when they meet halfway through, or the turn of a sonnet where thought, rhythm, and rhyme combine in one perfect word at the end of the stanza, Gödel's proof has that air of unconstructed inspiration, of having seen the whole in a glance.

Many years later Gödel described his thought process to Rudy Rucker as an almost mystical journey into the world of mathematical objects. "One must close off the other senses, for instance by lying down in a quiet place," Gödel explained, and then actively seek with the mind to *directly* perceive numbers, infinite sets, and the other "objective and absolute" inhabitants of the world of "pure abstract possibility." It was a mistake "to let everyday reality condition possibility, and only to imagine the combinings and permutations of physical objects." Mathematics, Gödel observed in one of his notebooks, demands a contradictory blend of humility and audacity. "It is fruitful to repeatedly reconsider seemingly insignificant and trivial theorems until one understands them perfectly," he wrote. But at the same time, a complete confidence of success when venturing into uncharted territory is essential in order to overcome the depressing feeling that a problem is "objectively pointless," and that requires holding fast to "the belief that what you are doing is the right thing, that the way you are approaching it is the right one, and that what you have done so far about it is right."[19]

Gödel's proof owes much of its otherworldliness to his use of a series of self-referential turns that suddenly, seemingly magically, pull the desired formula out of the hat at the very end. The German poet (and amateur mathematician) Hans Magnus Enzensberger, in his poem "Hommage à Gödel," likens Gödel's feat to the tale of Baron Münchhausen extricating the horse he was riding from the mud by pulling up on his own shock of hair:

> *Münchhausen's Theorem—Horse, Mud, and Hair—*
> *is enchanting, but don't forget*
> *Münchhausen was a liar.*
>
> *Gödel's Theorem at first glance*
> *seems rather nondescript, but consider:*
> *Gödel was right.*
>
> . . . .
>
> *So take these propositions in hand,*
> *And pull!*[20]

Gödel had done Robert Musil's metaphor of floating in air one better. His Incompleteness Theorem pulls itself into the air, rising above the edifice of mathematics by force of its own internal logic alone, to produce a formula that declares its own unprovability.

PROVABLY UNPROVABLE

Proving consistency was the heart and soul of Hilbert's program for a fundamental reason. As Gödel explained in an invited talk he gave to the Philosophical Society of New York University in 1934—though never published, and indeed still hardly known, it remains one of the best non-technical elucidations of his Incompleteness Theorem anywhere—if consistency goes, anything goes. If it were ever possible to derive both the statement *A* and its negation not-*A* within a mathematical system, Gödel observed, "it can easily be shown that *any* formula whatsoever would be provable"—even an outright absurdity like $0 = 1$.[21]

The problem in constructing a consistency proof within a mathematical system, though, comes down to chickens and eggs. "In order to be convinced by this supposed proof," Gödel pointed out, "we must

Fertig zum Umbrechen.
Dr Kurt Gödel
Wien VIII Josefstädterstr. 43.

Über formal unentscheidbare Sätze der Principia Mathematica und verwandter Systeme I).

Von Kurt Gödel in Wien.

1.

Die Entwicklung der Mathematik in der Richtung zu größerer Exaktheit hat bekanntlich dazu geführt, daß weite Gebiete von ihr formalisiert wurden, in der Art, daß das Beweisen nach einigen wenigen mechanischen Regeln vollzogen werden kann. Die umfassendsten derzeit aufgestellten formalen Systeme sind das System der Principia Mathematica (PM)²) einerseits, das Zermelo-Fraenkelsche (von J. v. Neumann weiter ausgebildete) Axiomensystem der Mengenlehre³) andererseits. Diese beiden Systeme sind so umfassend, daß alle heute in der Mathematik angewendeten Beweismethoden in ihnen formalisiert, d. h. auf einige wenige Axiome und Schlußregeln zurückgeführt sind. Es liegt daher die Vermutung nahe, daß diese Axiome und Schlußregeln dazu ausreichen, überhaupt jeden denkbaren Beweis zu führen. Im folgenden wird gezeigt, daß dies nicht der Fall ist, sondern daß es in den beiden angeführten Systemen sogar relativ einfache Probleme aus der Theorie der gewöhnlichen ganzen Zahlen gibt⁴), die sich aus den Axiomen nicht entscheiden lassen. Dieser Umstand liegt nicht etwa an der speziellen Natur der aufgestellten Systeme, sondern gilt für eine sehr weite Klasse formaler Systeme, zu denen insbesondere alle gehören, die aus den beiden angeführten durch Hinzufügung endlich vieler Axiome entstehen⁵), vorausgesetzt, daß durch die hinzugefügten Axiome keine falschen Sätze von der in Fußnote⁴) angegebenen Art beweisbar werden.

¹) Vgl. die im ▇▇▇ erschienene Zusammenfassung der Resultate dieser Arbeit.

²) Zu den Axiomen des Systems PM rechnen wir insbesondere auch: Das Unendlichkeitsaxiom (in der Form: es gibt genau abzählbar viele Individuen), das Reduzibilitäts- und das Auswahlaxiom (für alle Typen).

³) Vgl. A. Fraenkel, Zehn Vorlesungen über die Grundlegung der Mengenlehre, Wissensch. u. Hyp. Bd. XXXI. J. v. Neumann, Die Axiomatisierung der Mengenlehre. Math. Zeitschr. 27, 1928.

⁴) D. h. genauer, es gibt unentscheidbare Sätze, in denen außer den logischen Konstanten ¬ (nicht), ∨ (oder), (x) (für alle), = keine anderen Begriffe vorkommen als + (Addition), . (Multiplikation), beide bezogen auf natürliche Zahlen, wobei auch die Präfixe (x) sich nur auf natürliche Zahlen beziehen dürfen. In solchen Sätzen können also nur Zahlenvariable, niemals Funktionsvariable vorkommen.

⁵) Dabei werden in PM nur solche Axiome als verschieden gezählt, die aus einander nicht bloß durch Typenwechsel entstehen.

Gödel's marked-up page proof of his incompleteness paper

know that the axioms and rules of inference which we used will always lead to correct results. But if we knew this in advance, then no proof for freedom from contradiction is necessary."[22]

So "the chief point" in carrying through any proof of a formal system's freedom from contradiction is that "it must be conducted by perfectly unobjectionable methods": namely those that even the most irredentist of the intuitionists would accept.[23] Most of the modern mathematics that Hilbert wanted to save is built upon the use of infinite sets. But the whole idea behind Hilbert's notion of "proof theory" was that it should not assume "the validity of the very methods whose justification we are seeking," Gödel noted. "Now nobody has ever questioned seriously the consistency of *finite* math," he continued, so a consistency proof using only "finite methods, i.e., methods as are not based on the existence of infinite sets," and which are strictly constructive, would fit the bill.[24]

Gödel began by devising a method that would transform statements about the *process* of proof into statements about the properties of numbers. Doing so would allow him to write metamathematical assertions about the construction of proofs as simple arithmetical formulas, which could thus be expressed within the syntax of a mathematical system itself. Turning metamathematics into simple mathematical propositions was the first of his self-referential feats of sleight of hand.

The scheme he devised made it possible to convert any string of symbols used to express a mathematical or logical formula in *Principia Mathematica* into a single, unique integer. Russell and Whitehead had used a deliberately limited set of a dozen "primitive" symbols in setting up their system, and they showed that with these symbols any proposition of logic or number theory could be expressed, albeit in an often extraordinarily cumbersome fashion. Gödel first assigned a numerical value to each symbol. He used several different coding systems in his different presentations of the Incompleteness Theorem; in the series of lectures he gave in Princeton in 1934 he used the scheme:[25]

0	N	=	~	v	&	→	≡	∀	∃	∈	(	)
1	2	3	4	5	6	7	8	9	10	11	12	13

(For definitions of these symbols and a fuller explanation of Gödel's proof, see appendix.)

The particularly clever part of the coding scheme is that it is reversible. By using powers of prime factors to generate the integer representing a string of symbols in a complete formula such as "$x = 0$," the resulting formula number, its "f-number" Gödel called it, can be "decoded" through a straightforward arithmetic process back to the statement $x = 0$.

Not only that, but by examining the f-number's arithmetic properties, one can determine whether the expression it stands for possesses certain features. For example, an expression with an f-number divisible by 3^3, or 27, but not by any higher powers of 3, will always have an equal sign as its second term (as in the expressions $x = 0$ or $0 = 0$). Similar numerical tests can establish if an f-number represents an expression with, say, two left parentheses but no right parentheses, or any other such rule of syntax. So that means it is possible to determine, *just* from an f-number's arithmetic properties, whether the string of symbols it stands for forms a meaningful mathematical or logical statement such as $x = 3$ or $\sim (A) \rightarrow B$, or a meaningless string of gibberish like $=((\rightarrow \&$. As Gödel noted, this says nothing about whether the statement is *true*, just whether the expression is constructed according to the formal rules of the system.

Likewise, examining the arithmetic relationships between the f-numbers of two or more expressions can reveal whether one is a valid inference of the other according to the system's permitted rules of inference. Like any other formal system, *Principia Mathematica* is built up of both axioms, self-evident starting points like the assertion $0 = 0$, and rules of inference that permit one to draw valid conclusions from one statement to another, such as the rule that $A = B$ implies $B = A$, or that A and $A \rightarrow B$ together imply B. A numerical formula specifying all of the valid rules of inference would be mind-boggling in its size, but as there are a finite number of them there is no reason in principle it could not be done. The key point is that by this means Gödel was able to express the *logical* relation between two formulas as a *numerical* relation between two numbers.

Gödel's final preparatory step was to extend his coding system to complete proofs. A proof is a sequence of formulas, so a complete proof can be represented by a sequence of numbers. Combining all the numbers in a proof sequence into a single immensely large integer, Gödel produced an f-number that stands for an entire proof.

But again, Gödel emphasized, the properties that determine whether such a proof number represents a valid proof are "purely *arithmetic*."[26] The first number of the sequence would have to be the formula number for one of the axioms; the last number would have to be the formula number for the proposition being proved; and each number along the way, in what Gödel called the "chain of inference," would have to obey the numerical rules that specify a valid application of a rule of inference from a previous formula in the chain.

So, by Gödel's coding scheme, *proofs* in number theory can themselves be represented as *propositions* of number theory. He was thus able finally to construct an arithmetic expression which states, "there is no value of x such that x is the proof number for formula number z," a function he designated using an abbreviation of the German word "provable," *beweisbar*, with a bar over the top signifying the mathematical shorthand for negation:

$$\overline{Bew}\,(z)$$

The numerical formula that $\overline{Bew}\,(z)$ represents, as long and complicated as it would be, is nonetheless a purely arithmetical statement made up of nothing but strings of the primitive symbols of *Principia Mathematica*—parentheses, equal signs, x's, and so on. That means the statement $\overline{Bew}\,(z)$ itself has a formula number, too, just like any other mathematical formula within the system.

As Gödel then observed, with considerable understatement, "This leads to interesting results."[27] He delivered the coup de grace in an extraordinary feat of mathematical legerdemain through which he proved that it is always possible to find some value of z, which he called g, such that the formula number for the resulting version of the formula $\overline{Bew}\,(z)$ is none other than . . . g itself. In other words, he had succeeded in constructing a proposition G,

$$\overline{Bew}\,(z) \qquad\qquad \textit{proposition G}$$

which asserts that there is no possible series of proof steps, starting from the axioms of the system and proceeding by its valid laws of inference, by which formula number g can be derived. But formula number g is that proposition G itself. G in other words states: "G is unprovable."

As Gödel observed, this means that G is *formally undecidable*: it is impossible to construct within the system a proof of either G or of its negation, not-G.

Although Gödel in the introduction to his first full publication of the Incompleteness Theorem noted its similarities to the Liar's Paradox, he had in actual fact brilliantly sidestepped that pitfall by substituting the concept of *provability* for the concept of *truth*. While a pair of statements of the form A and not-A cannot both be false, they can both be unprovable, which was exactly what Gödel had demonstrated in the case of his formula G.[28]

Whether true or false, a statement that says, "This statement is unprovable," is trouble. If true, it is itself an instance of a true but unprovable proposition. But if it is false, that means it *can* be derived from the axioms of the system, and is thus an instance of a false but provable statement, which is arguably even worse.

Gödel slammed shut the final hope of salvation by showing that *any* formal system which contains arithmetic, not just *Principia Mathematica*, will suffer from the identical flaw. Nor is there a way out by adding G to the axioms of the system—in other words, just declaring it to be a given that we need not establish by formal proof; Gödel showed that that would just generate yet *another* new undecidable statement somewhere in the system.

BITTER FRUIT AND SOUR GRAPES

Two days after Gödel dropped his casual bombshell into Hilbert's program at Königsberg, he was in the audience for Hilbert's speech to the Society of German Natural Scientists and Physicians. It was something in the way of a farewell address. Hilbert was sixty-eight years old, and a ceremony was planned by the city of Königsberg to name him an "honorary citizen" of the town of his birth. His speech was to be his formal reply to the honor, and perhaps his last great chance to give his vision for the future of mathematics.

Afterward, Hilbert was escorted to a nearby radio studio to record a broadcast in which he would repeat the concluding remarks of his speech. A 45-rpm recording survives, and in it Hilbert ends by returning to the theme of his famous 1900 challenge. For those who pursue truth through science, he insisted, "there is no *ignorabimus*." That word was an unmistakable allusion to the address given to the very same society in 1872 by the rector of Berlin University, the physiologist Emil Du Bois-Reymond, who had argued that all scientific knowledge is bounded by the two limits of mind and matter, and therefore there will always be truths beyond the reach of man. *Ignorabimus*, the word that Du Bois-Reymond had ended his speech with, came from the Latin maxim *Ignoramus et ignorabimus*: "We do not know, and we will not know." In his concluding line Hilbert turned the phrase around: "We *must* know," he said. "We *will* know."[29] When he died in 1943 the words were engraved on his tombstone at Göttingen:

WIR MÜSSEN WISSEN

WIR WERDEN WISSEN

Had Hilbert been in the audience for Gödel's talk two days earlier, he might have felt compelled to add an asterisk.

News of Gödel's result spread quickly, and even if not everyone immediately understood his proof there was no doubt something big had happened. "What is going on with Herr Gödel?" the philosopher Heinrich Scholz wrote to Carnap from Münster. "I hear all kinds of exciting things but I cannot make out what it is all about." From Paris his friend Natkin sent his congratulations, having heard all about it from Feigl:

During Herbert's last visit he told me much about you, and unjustifiably I am terribly proud. . . . So you have proved that Hilbert's axiom system has unsolvable problems—that is no small thing.[30]

One of the first to completely grasp the earthshaking implications—
that Gödel had torpedoed the very possibility of *ever* placing mathematics on a foundation of logical consistency—was John von Neumann,
who cornered him immediately after the discussion at Königsberg
and closely interrogated him about his result. Von Neumann himself
had been working to prove the consistency of mathematics and in fact
was close on Gödel's trail. He was formidable competition. Stories of
his lightning-quick facility for mental calculations and feats of pho-
tographic memory abounded. He had been able to divide eight-digit
numbers in his head at age six and as a child joked with his father in
classical Greek, and throughout his life he could recite entire books
verbatim after a single reading, providing a running translation into
another language, if requested, at no diminution in speed. Once his
colleague Herman Goldstine tried to test him by asking him how
Dickens's *Tale of Two Cities* began. He was still going fifteen minutes
later without pause when Goldstine finally stopped him. His memory
equipped him with an "unparalleled storehouse" of anecdotes, stories,
and humorous and dirty limericks, Goldstine recalled, which he would
bring out to enliven otherwise dull social and scientific discussions.[31]

Von Neumann had caused Gödel a momentary panic when he
wrote on November 20, about two months after the Königsberg meeting, to congratulate him on his theorem ("the greatest logical discovery
in a long time") and to inform him of a "remarkable" result he had
just derived from it. Von Neumann then outlined a simple proof which
showed that, as a consequence of Gödel's theorem, it is impossible ever
to prove the consistency of a consistent system.[32]

Gödel immediately wrote back to say he had already found that
result himself:

> Unfortunately I have to inform you that I have for about three
> months now already been in possession of the result you commu-
> nicated. . . . The reason why I didn't make any presentation of the
> above result is that the precise proof is not suited to oral communi-
> cations and an approximate indication could easily arouse doubts
> about the correctness. As concerns the publication of this matter,

there will be given only a shorter sketch of the proof of impossibility of freedom from contradiction in the *Monatsheft* that will appear in the beginning of 1931 (the main part of this treatise will be filled with the proof of the existence of undecidable sentences).[33]

To substantiate his claim of priority Gödel enclosed an offprint of an abstract he had sent to the Vienna Academy of Sciences in September. Von Neumann promptly replied, "As you have established the theorem about the unprovability of consistency as a natural continuation and extension of your earlier results, I of course will not publish on it."[34]

Gödel was not exaggerating in claiming to have beaten von Neumann to the punch, but he nonetheless hastily rewrote the draft of his *Monatshefte* paper to add a section outlining this second discovery, what is now generally called Gödel's Second Incompleteness Theorem, and he added a roman numeral "I" to the title of his paper to imply that there was a second paper ready to be published, offering the full proof. In the end that would prove unnecessary. No one doubted either his priority or the substance of the Second Theorem.[35]

Von Neumann's deep esteem for Gödel's accomplishment tempered any personal regrets. He acknowledged years later to his friend the mathematician Stanislaw Ulam that he was disappointed at not having come up with the idea of undecidability himself. But he admitted to Herman Goldstine that he had been pursuing the wrong track in any case, having at one point come close—or so he believed—to proving the *consistency* of mathematics, the exact opposite of Gödel's result. While working on the problem he had, on two successive nights, dreamt of a possible solution to an obstacle he had encountered in his proof. Each time, he had excitedly gotten out of bed and gone to his desk to try out the idea; each time, it got him a bit further along, but the snag remained. "How lucky mathematics is," he remarked to Goldstine, "that I didn't dream the third night!"[36]

As von Neumann would say on the occasion of presenting Gödel the Einstein Award two decades later, his achievement "is singular and monumental—indeed it is more than a monument, it is a landmark which will remain visible far in space and time." In the fall of 1930 von

Neumann was giving a lecture course in Berlin on Hilbert's program to rescue classical mathematics with finite methods when he received from Gödel an advance copy of his *Monatshefte* paper. "I recall it was a dramatic incident," said Gustav Hempel, who was attending the class. "It was in the middle of the semester, in late autumn, a beautiful day as I recall, and Neumann came right in and said he had received this news from a young mathematician or logician in Vienna, and it showed that Hilbert's goals could not be achieved, and so he would stop discussing this topic, and devoted the rest of the semester to a discussion of Gödel's ideas."[37]

Menger, who was on leave in America for the academic year, learned of Gödel's discovery in late January 1931 while lecturing at the Rice Institute in Texas. Gödel had given a talk at the Mathematical Colloquium two weeks earlier about his result, and Menger received a letter from Vienna acquainting him with the news. At the end of Gödel's presentation, recalled the mathematician Franz Alt, there had been a stunned silence—broken by the remark from one of the participants, "That is very interesting. You should publish that."[38]

Menger, like von Neumann, immediately interrupted his course to explain Gödel's work, bringing the first news of it to America. On his return to Vienna, he was the first to present an explanation of the Incompleteness Theorem and its significance to the general public, in his presentation on "The New Logic" as part of the opera-ticket-priced series of public lectures he organized in 1932.[39]

Gödel's proof was met with incomprehension and hostility, however, by several in Hilbert's camp. Hilbert himself, according to his assistant and collaborator Paul Bernays, was "somewhat angry" when he first learned of Gödel's proof.[40] Though Hilbert later included the full proof in the second volume of his *Die Grundlagen der Mathematik* in 1939 and acknowledged its correctness, he refused to concede that it meant the end of the formalist program or the goal of demonstrating the consistency of arithmetic:

With regard to this goal, I would like to emphasize that the temporary opinion that certain recent results of Gödel imply the impracticability of my theory of proof is erroneous. Indeed, that result only shows that one has to use the finite viewpoint in a sharper way for the further proofs of freedom from contradiction than is necessary when considering the elementary formalisms.[41]

Bernays for his part, while complimentary, at first seemed quite unable to grasp Gödel's distinction between truth and provability. Ernst Zermelo, a German mathematician who had contributed major advances to logic and set theory, did not grasp it either, but he chose to go on the warpath and claimed to have found a mistake in Gödel's proof. At a meeting of the German Mathematical Society in Bad Elster in September 1931, he made it clear he had no wish to meet this young upstart, resisting Olga Taussky's urging that they sit down and talk to each other. When a small group proposed walking up a nearby hill for lunch, Zermelo objected that if Gödel came along there would not be enough food, that the climb would fatigue him, and then pointing to a person he (wrongly) assumed was Gödel, told Taussky he had no desire to speak to anybody with such a stupid face. But within seconds of their finally, somehow being introduced, "a miracle happened," Taussky wrote. "The two scholars were engaged in deep contemplations and Zermelo walked up the mountain without even knowing that he did it."[42]

Still, in their subsequent correspondence, Zermelo obstinately persisted in dismissing Gödel's proof on the grounds that his function *Bew* would simply yield a paradox like Russell's if it were changed to an assertion of truth rather than of provability—which of course was exactly why Gödel had not done so. For years, similar attacks cropped up from time to time, most of which zeroed in on the loose analogy Gödel had drawn in his introduction between his proof and the logical paradoxes, to insist that all Gödel had done was to manufacture a self-contradictory definition. But these were potshots from skirmishers. The fact was that Gödel had carried the field with remarkable swiftness.[43]

The sourest grapes came from Brouwer, who contented himself with grouchily remarking that he was the only person who was unsurprised by Gödel's proof, and that it did not change his views anyway—since he did not believe that mathematics could or should be contained within a formal system.[44]

The saddest footnote involved the American logician Emil L. Post, who believed he had "anticipated" Gödel's result in work he had been doing on undecidable propositions since the early 1920s. After they met, for the first time, in New York in 1938, Post wrote a despondent and gracious letter in which he apologized for having been overwrought by their meeting. "But for fifteen years," he explained, "I had carried around the thought of astounding the mathematical world with my unorthodox ideas, and meeting the man chiefly responsible for the vanishing of that dream rather carried me away." He ended by saying, "As for any claims I might make perhaps the best I can say is that I would have *proved* Gödel's Theorem in 1921—had I been Gödel."[45]

Still, Post was able to contribute an important extension of Gödel's theorem which, along with a near simultaneous paper by Alan Turing, provided a rigorous definition of a formal system as a series of basic, mechanical computational steps—the conceptual model for a computer that would come to be known as the Turing Machine, and which would lay the foundation of modern computer science. Gödel and Turing never met, but each recognized the great significance of the other's work for the newly emerging field of computing. Gödel in turn credited Post's and Turing's "precise and unquestionably adequate definition of the general concept of a formal system" for establishing the complete generality of his proof in "*every* consistent formal system containing a certain amount of finitary number theory."[46] Post and Turing independently had come up with the idea of a conceptual machine that could perform any calculation by employing a program with just a few primitive binary operations: marking a square on a strip of paper, erasing the mark, moving one square to the left or right, and reading whether the current square is marked or not. Their profound insight was not only that a formal mathematical system could be defined by such a mechanically reproducible procedure, but that Gödel's undecidable propositions were exactly equivalent to a program running on such a machine that never reached a result, but kept running forever.[47]

Gödel from the start was thinking, too, about the implications his theorem held for mathematics and philosophy as a whole. Whatever gloomy thoughts he might later entertain that he had only shown what "cannot be done," Gödel then and later drew an encouragingly optimistic lesson from his astonishing result. It meant, he said, that mathematics was "inexhaustible": there would always be new mathematical truths for the human mind to discover. If some could not be derived within the confines of a given formal system, that simply meant that human intuition was a more magnificent thing than what any machine could duplicate.

As far as he was concerned his result did not upend Hilbert's program. It just meant that human ingenuity would be required to build new paths to the truths that were out there, waiting to be found.

BABBITTS ON PORCHES

In June 1932 Gödel submitted his Incompleteness Theorem for his Habilitation thesis, and this time his adviser Hans Hahn did not stint praise. "It is a scientific achievement of the first order," he wrote, "which has generated the highest interest in all expert circles and, as can be predicted with certainty, will take its place in mathematical history."[48] There was an informal rule at the University of Vienna that an applicant for a Habilitation in mathematics had to wait at least four years from the awarding of his doctorate before seeking this next rung of promotion on the academic ladder, but the rule was waived in Gödel's exceptional case.

Still, with all of the bureaucratic hurdles to be cleared it was nearly three years from the time he received his PhD before his Habilitation was formally approved. Besides his thesis, an applicant had to submit three possible topics for a probationary lecture which would be evaluated, pass an oral examination during which he could be asked questions on any subject in his field, and throughout the process receive favorable votes by an examining committee and the full College of Professors, who among other things had to certify that "the personality of the candidate gave no grounds for doubt as to his suitability for a university teaching position." The final vote of the faculty was 42 yes,

1 no, and 1 abstention. The name of the anonymous objector to Kurt Gödel's promotion is lost to history.[49]

Passing the Habilitation conferred the title of *Privatdozent*—"private lecturer"—which was a license to teach at the university but not a license to make a living at it. The rank came with no position or salary. A dozent's only income was a small fee, the *Kollegiengeld*, paid directly by each student enrolled in his classes. Dozents were required to offer a minimum of two hours of lectures a week every fourth semester, but even lecturers far more popular with their students and more diligent in their duties than Gödel ever was made hardly enough to support themselves from teaching alone. In his entire time in Vienna Gödel offered only three courses: foundations of arithmetic in 1933, select topics in mathematical logic in 1935, and axiomatic set theory in 1937, thereby just meeting the minimum requirement. One of the students in his logic class—the number theorist Edmund Hlawka—recalled that Gödel lec-

Paltry lecture fees from 1937

tured at breakneck speed, facing the blackboard the entire time, with his back to the classroom. "I can honestly say I soon understood nothing at all," Hlawka said. By the end of the semester, the class, which had begun with a full lecture room, had only a single student left. (That was the Polish mathematician Andrzej Mostowski, who became a distinguished logician.) A receipt for one semester's work that Gödel carefully filed away records his payment in the sum of 2 schillings 90 groschen, enough to buy two beers.[50]

Gödel's friends tried to help him out by sending small mathematical editing jobs his way. In the summer of 1932 Menger offered him 250 schillings from a grant he had received to publish a geometry textbook, at the same time enclosing a Charlie Chan mystery so he could improve his English. (Gödel confessed a couple of months later that he had not made much headway on the novel: "It is very laborious for me to read.") At Menger's request he also helped out regularly with the journal of his colloquium, *Ergebnisse eines Mathematischen Kolloquium*, and contributed book reviews and short articles to other mathematical journals. A list of his income from October 1932 to July 1933 includes 24 schillings 70 groschen in *Kollegiengeld*, 915 schillings from his father's pension from the Redlich Fabrik, and other odds and ends for a grand total of 4,286 schillings 70 groschen, or about $700 at the time. Far outspending his income, he was rapidly using up the inheritance that was making his comfortable lifestyle for the time being possible.[51]

Outside of those far from full-time occupations, he continued his mathematical work, still rising late in the morning and staying up into the night, met his friends at coffeehouses, went to the theater and movies and concerts with his mother and brother and had long discussions afterward about what they had seen, went on hikes and other family excursions in the hills around Vienna on weekends and spent summer vacations near the Rax Mountain, about fifty miles south of the city. A pleasant photograph of him from this time shows him in plus-fours and long wool hiking socks, wearing a soft cap and carrying a walking stick on a country lane with the mountains behind.

Rudi Gödel had completed his medical studies and was practicing as a radiologist in one of the leading clinics in Vienna by this time,

Right: Hiking near the Rax Mountain

Below: Josefstädter Straße; Gödel shared an apartment in the building on the right with his mother and brother from 1929 to 1937

and they were all still living amicably together at the apartment on Josefstädter Straße, the only small source of friction being Aunt Anna, whom Rudi described as "a thoroughly good-natured, nice lady, who only through her pessimism and extreme anxiety about all things of life sometimes got on my mother's nerves."[52]

Given the near-impossibility of finding a permanent academic position in Austria or Germany following the worldwide recession of 1930, nearly everyone in Gödel's circle of extraordinarily able mathematicians and philosophers was looking for opportunities elsewhere. The Vienna Circle was already beginning to break up. Feigl was the first to depart, spending a year at Harvard on a Rockefeller Foundation fellowship in 1930 before taking a position the following year at the University of Iowa.

Two months after his arrival in the distinctly uncosmopolitan metropolis of Iowa City (pop. 15,000), he sent his friend Gödel a mordant description of life in the American heartland, underscoring just what a wrenching adjustment anyone from Vienna faced transporting himself to a small American college town. The American Dream was not all it was cracked up to be. Train fares and the cost of domestic help were unbelievably high; the absence of sidewalks outside of town made going for a walk impossible ("Americans as a rule do not go for walks, they seem to think that dashing around in their cars on Sundays is sufficient recreation"); the chlorine-disinfected river water made him long for Vienna's famous Alpine-fed tap water; and the identical one-story homes, "with mostly tasteless porches on which the Babbitts in rocking chairs read newspapers," and their gloomy overheated living rooms and tiny bathrooms, were depressing. "To our disappointment, there is not much to see of the vaunted American comfort."

Everything at the university was hopelessly overorganized and overstaffed. But the greatest disappointment was the dreary and dull society of congenitally immature Americans, which Feigl nonetheless allowed was offset to some extent by their fundamental decency:

> Overall, the internal connections of the faculty are stronger than ours; one sees each other at all kinds of meetings and also privately.

One is allowed less intrigue and spitefulness than with us. You let your fellow man live and there is amazingly little personal intrigue. There is much socializing, but due to its superficiality terribly boring. We try to get through with a minimum, which is not at all easy: Maria is constantly visited by the professors' wives; the rule demands that you reply to these visits, from which follow dinner invitations, soon one is inevitably forced into an endless chain of boredom. Luckily the men are not so terribly bad as the wives, who constantly get together in clubs and generally rule their lives as a kind of higher power. Maria has so far narrowly avoided the clubs, and I hope very much she continues to succeed. . . .

For a doctor of philosophy one must go to school for at least 20 years, but people at 30 years are perhaps as mature as our 18-year-olds. It is the immaturity and naiveté that makes the people so uninteresting, we think. . . . Our impression is that nervous and neurotic characters are almost completely absent here, at least nothing is outwardly shown. People are robust, healthy, self-assured, the women almost all pretty but not one beautiful, all dressed fashionably and well but not distinctively and not truly elegantly. It is not so much from a feeling of inferiority as from an aspiring childishness that Americans in all their actions and speech (discreetly but clearly) keep pointing out, "We can do that too, we also have culture."

Feigl was desperate to get a position in New York where there would be "more excitement, more opportunity for music, and interesting contacts," he said. "It seems you are somewhat away from real life here."[53]

Back in Vienna, real life was taking a darker turn that would further hasten the rush to the exits. Two months after coming to power in Germany in January 1933, Adolf Hitler issued the Law for the Restoration of the Civil Service, which required all "non-Aryan" civil servants to "retire." That included university professors, and it was the start of what would become the largest brain drain in history as thousands of Germany's—and later Austria's—Jewish physicists, mathematicians, and other scientists fled the Nazis, mostly to America and Britain.

One of the first to leave was Albert Einstein, who accepted a position at the newly founded Institute for Advanced Study in Princeton. "As long as I have any choice in the matter, I shall live only in a country where civil liberty, tolerance, and equality of all citizens before the law prevail," Einstein told a reporter from the *New York World Telegram*. "These conditions do not exist in Germany at the present time." The outraged Nazi government ordered the Prussian Academy of Sciences to expel him for his "foreign agitation." Einstein took satisfaction in resigning before they could act.

In Austria, too, the National Socialists were rapidly gaining adherents, and foreign journalists were reporting that the party might receive as much as 50 percent of the vote in the next elections. On March 4, 1933, Austria's Christian–Social chancellor, Engelbert Dollfuss, invoked a wartime emergency law still on the books to suspend parliament and sent police to bar members from entering. Hoping to erect a bulwark for Austrian independence, Dollfuss imposed his own clerical–fascist counterpart to Nazism, complete with its home-grown version of the fascist cross—the *Kruckenkreuz*, the Jerusalem cross, in place of the Nazis' swastika, or *Hakenkreuz*—as the emblem of his new, papally backed, "Christian and corporatist" state. Instead of tumultuous political parties and popular elections, the parliament would henceforth be selected by councils representing the different constituencies of the state—farmers, workers, business, church—all under the aegis of a single supra-partisan organization, the Vaterländische Front. While most of Gödel's circle were willing to accept Austrian fascism as a far better alternative than Nazism—and Gödel like many other erstwhile liberals dutifully enrolled as a member of the slightly comic-opera Vaterländische Front as a necessity for retaining a public position—the rapid changes in Austria's political landscape in the spring of 1933 added to daily life in Vienna a sense of deep tension and apprehension of what might come next.[54]

A few months later Menger confided to an American mathematical acquaintance, Oswald Veblen, the almost unbearable strain of daily events. Menger was visiting Geneva, where at last he felt able to speak freely:

What I could not write you from Vienna is a description of the situation there. You know how fond I am of Vienna and how many things I started there in the intention of staying there still a good many years. But the moment has come when I am forced to say: I hardly can stand it longer. First of all the situation at the university is as unpleasant as possible. Whereas I still don't believe that Austria has more than 45% Nazis, the percentage at the university is certainly 75% and among the mathematicians I have to do with, not far from 100%. So unpleasant this is, also for me personally, it is for me personally not the very worst circumstance, since the political activity of the University-Nazis consists in keeping away non-Nazis from administrative business from which I am only glad to stay away. It is the general political atmosphere whose tensions and impending dangers ever since two years an always increasing strain of nerves which begins to become unsupportable. Not that I had particularly weak nerves. But you simply *cannot* possibly find the concentration necessary for research if you read twice a day things in the newspapers (whose reading is indispensable for a man in a public position) which touch the basis of the civilization of your country as well as your personal existence. And periods of this strain become longer and longer and more and more frequent.[55]

He ended by asking if Veblen might help him find some position in America, but begged him to be circumspect in his reply so as not arouse the attention of anyone who might see or open his letter, and not to repeat anything he had said about the political situation to any European.

The previous summer, Menger had seen an opportunity to do his friend Gödel a good turn on the same matter. Veblen had telephoned in June 1932 to say he would be passing through Vienna, and asked if they might meet. Veblen in fact was on a months-long talent-spotting trip across Europe, looking for first-rate theoreticians to help get the new Institute for Advanced Study off the ground. Menger asked Gödel to prepare a lecture on a result he had just mentioned: that by means of an intriguing reinterpretation of basic concepts, all of classical arithmetic and number theory is derivable in intuitionist mathematics. Veblen was invited to tea with Professor Hahn on June 29 and the next day at 1 p.m. came to the Mathematics Institute to hear Gödel's talk.[56]

Duly impressed, Veblen wrote Menger on his return, "I hope that the new Institute makes it possible to carry out the plan which you and I discussed with regard to Godel. We might offer him a position in the Institute for one year at a salary of something like $2500. . . . Since the Institute is purely a research institute, he would not be required to give any lectures. If, however, he wished to give a course of lectures or a seminar, there would no doubt be students in the University who would like to hear it. . . . I think the offer would simply give Mr. Godel a free year in which to pursue his research in pleasant surroundings."[57]

Because he would need to remain in Vienna to deal with his pending Habilitation requirements, Gödel delayed replying for several months. Veblen finally cabled. In January 1933 Gödel replied, "ACCEPT GLADLY STOP LETTER FOLLOWS GOEDEL." He explained in the following letter that there was no need to advance him his travel expenses, as Veblen had offered, as he had sufficient funds in hand, though did ask if the salary could be guaranteed against a drop in the value of the U.S. dollar, a request which Veblen tactfully deflected.[58]

A group of friends, including Olga Taussky, came to see him off at the Westbahnhof, where he boarded the wagon-lits of the Orient Express for the first leg of his journey. "A fine looking gentleman, presumably his Doctor brother, stood apart from us and moved away as the train started, while we others waved a little longer," Taussky remembered.[59]

From Feigl in Iowa City came a jocular note of congratulation. "So you too, my son—like Einstein and all other celebrities, at last could not help it—and had to come across the great waters. Quite right—it will probably lead to a permanent position in the end, and the Germans and Austrians will have once again lost a (this time racially pure (?)) scholar."[60]

· 6 ·

The Scholar's Paradise

A COUNTRY CLUB FOR MATH

As Olga Taussky later learned, that farewell at the Westbahnhof had not been Gödel's actual departure. Intending to sail on the Cunard liner RMS *Berengaria*, which departed Cherbourg for New York on September 23, 1933, he had not gotten far when he totally lost his nerve. Taking his temperature and finding it slightly elevated, he turned around and headed straight back to Vienna. It should have been a warning of more serious troubles ahead, but his mother and brother convinced him to try again a week later.[1]

This time he managed to make it aboard the RMS *Aquitania*, sailing from Cherbourg on September 30. On the list of arriving alien passengers at the Port of New York, his occupation was given as "teacher"; height, five feet seven inches; hair, brown; eyes, brown; complexion, dark; health, good; not an anarchist or polygamist. He was met at the dock by Edgar Bamberger, a member of the board of the Institute for Advanced Study and nephew of its wealthy founding patron. He arrived four days too late to be present for the Institute's official opening on October 2.[2]

The Institute for Advanced Study was the brainchild of Abraham Flexner, an indefatigable crusader for the reform of higher education in America. Flexner had made his reputation twenty years earlier with a damning exposé of shoddy medical schools. Born in Louisville, Kentucky, in 1866, the seventh of nine children of an immigrant Jewish pushcart peddler from Bohemia, Flexner graduated from Johns Hopkins University, where his tuition was paid by his eldest brother. Returning to Louisville, he taught Latin in the local high school, where he proved his fearless commitment to standards by flunking the entire class his first year. When the Carnegie Foundation for the Advancement of Teaching commissioned him in 1910 to examine the state of American medical education, he personally visited every one of the 153 medical colleges in the United States and Canada, and found that all but a handful were little more than diploma mills, with no admission standards, no laboratories, no practical training, and no graduation requirements other than handing over a check. The report earned him libel suits, death threats, and a national reputation overnight. As a direct result of his investigation, two-thirds of the colleges closed and sweeping reforms were instituted to ensure that rigorous and scientifically grounded training of doctors became the norm.[3]

John von Neumann's second wife, Klári, later described Flexner as "a small hawk-like wiry man with a wonderful twinkle in his eye and a front of obviously false modesty that immediately made you suspect the strength and power, the cunning and cleverness that were hidden behind that delightful sense of humor."[4] In 1929, after fifteen years at the Rockefeller Foundation, where he efficiently gave away $50 million of John D. Rockefeller Sr.'s money in grants to aid development of medical education, he turned his attention to the state of American higher education as a whole. Despairing that any American institution even deserved the name "university," he proposed a radical remedy. Rather than communities of scholars engaged in advanced research and teaching, he scathingly concluded, American universities were aimless agglomerations of "formless and incongruous activities," their graduate programs usually a poorly funded appendage and afterthought. With a "lack of any general respect for intellectual standards, the intrusion of politics here and of religion somewhere else, the absurd notion that ideals are 'aristocratic,' while a free-for-all scramble which distresses the able and intelligent is 'democratic,' " the only solution Flexner could see was to abandon any hope of reforming existing institutions, and to create a new, ideal one from scratch.[5]

That ideal place, he imagined,

> should be open to persons, competent and cultivated, who do not need and would abhor spoon-feeding. . . . It should furnish simple surroundings—books, laboratories, and above all tranquility—absence of distraction either by worldly concerns or by parental responsibility for an immature student body. Provision should be made for the amenities of life in the institution and in the private life of the staff. It need not be complete or symmetrical: if a chair could not be admirably filled, it should be left vacant. . . . It would be small . . . but its propulsive power would be momentous.[6]

The faculty of this intellectual Eden, as he later explained, would "consist exclusively of men and women of the highest standing in their respective fields of learning, attracted to this institution through its appeal as an opportunity for the serious pursuit of advanced study and because of the detachment it is hoped to secure from outside distractions."[7] He had been trying for years to turn his alma mater, Johns Hopkins, into a postgraduate-only institution that might fulfill this vision, but the idea had met the predictable resistance. And then one day in late December 1929 two men came to call on Flexner at the offices of the Rockefeller Institute for Medical Research in Princeton. They were the lawyer and business adviser of a wealthy brother and sister, Louis Bamberger and Caroline Fuld, who were looking to find a use for $10 million.[8]

Bamberger was the owner of the leading department store in Newark, New Jersey, which had grown into a 3,500-employee, $30-million-a-year business on his philosophy of treating both customers and employees well. The price of all merchandise was clearly marked on tags; refunds and exchanges were offered without question; salespeople were well paid (a sign in the store's restaurant asked patrons not to tip the waitresses, as they were adequately compensated with full salary). At age seventy-four, looking to retire, he had sold the entire business to Macy's Co. just six weeks before the stock market crash of October 1929. He and his sister walked away with $11 million in cash, and an additional 69,200 shares of Macy's stock (which lost half their value in the following two months).[9]

A shy and private man, Bamberger had never married, and his sister had no children. After giving away $1 million to the firm's 225 employees who had worked for them fifteen years or longer, they were still left a sizable fortune to dispose of. In the words of an internal history of the Institute, "They regarded their wealth not only as a just reward for the many years of faithful attention to the exacting business of serving the people of Newark . . . but also as a trust to be devoted to the welfare of their fellow citizens." Their first idea was to found a medical school that would give preferential admission to Jewish students. That was what had led them to seek Flexner's advice. He quickly disabused them of the wisdom of that idea, insisting that Jews did not face discrimination at medical schools, and pointing out moreover that any medical school needed to operate a good teaching hospital and be part of a strong university. Newark was also too close to New York City to be able to compete for staff and students.[10]

But on his desk were the galley proofs of the book he was about to publish decrying the state of the American university. "Have you ever dreamed a dream?" he innocently asked his visitors.[11]

Over the next two years he skillfully cajoled his newfound benefactors into realizing his plan for creating a pure "society of scholars" free to pursue what he, in a *Harper's* magazine article a few years later, famously extolled as the "usefulness of useless knowledge." When the Bambergers kept insisting that the new Institute would have to be in or near the city of Newark, Flexner kept adroitly sidestepping the issue, substituting "the State of New Jersey" for "the vicinity of Newark" in draft documents until finally he got his way. The fact was he had already made up his mind that Princeton was the place. Landing Albert Einstein removed any residual ill feelings: the Bambergers were ecstatic that the most famous scientist in the world would be joining their new Institute as one of its first five professors. *"Ich bin Feuer und Flamme dafür*"—"I am fire and flame for it"—Einstein had told Flexner when he received his astonishingly generous offer.[12]

Three of the five were there in Princeton on the day the Institute was officially inaugurated at a low-key meeting called by Flexner in his new office on the Princeton campus. All had ties to the university: Oswald Veblen and James Alexander, both of whom had been full professors in the mathematics department, and John von Neumann, who had been a part-time visiting professor there for the last two years. Einstein was to arrive from Europe a few weeks later. So was Hermann Weyl, who after months of indecision had accepted Flexner's offer and resigned his post at Göttingen, fearing for the future of his Jewish wife in Nazi Germany.

The decision to begin the Institute with a cadre solely of mathematicians was another of Flexner's calculating moves. He wanted to make an instant splash, and he knew that in mathematics there was broad consensus on who the leaders of the field were. Moreover, as he explained to the trustees, "Mathematics is singularly well suited to our beginning. Mathematicians deal with intellectual concepts which they follow out for their own sake, but they stimulate scientists, philosophers, economists, poets, musicians, though without being at all conscious of any need or responsibility to do so." Moreover, no elaborate labs or facilities would need to be constructed: "It requires little—a few men, a few students, a few rooms, books, blackboard, chalk, paper, and pencils."[13]

Admitting that "mathematicians, like cows in the dark, all look alike to me," Flexner tasked Veblen with the job of assembling his blue-ribbon herd.[14] Veblen had long had his own dream of building in Princeton a mathematical utopia. Like Flexner he also had a considerable talent for lavishly spending other people's money in the cause of intellectual endeavor. A nephew of the famous sociologist Thorstein Veblen, he was the grandson of Norwegian immigrants who had cleared the land and built a house and barns with their own hands on a series of homesteads in the upper Midwest, and then sent all nine of their children to college. At the University of Iowa Oswald Veblen had won two prizes: one in mathematics, the other in sharpshooting. Tall and lean, he was a lifelong outdoorsman, outwardly shy with a stut-

Oswald Veblen

ter but, as Klári von Neumann recalled, "a formidable opponent when anybody crossed his path." Herman Goldstine described him as "the kind of guy who would keep dripping water on the stone until finally it eroded."[15]

Veblen had worked assiduously to make Princeton a leading center of mathematics, and like Flexner he had come to believe that splendid isolation from worldly cares was the key to mathematical contemplation. He personally oversaw the construction of the university's magnificent new mathematics building, Fine Hall, which opened its doors in 1931. The building was named after Dean Henry Fine, the first to begin building up Princeton's mathematics programs. A wealthy fellow alumnus and friend of Fine's, a Chicago lawyer named Thomas Davies Jones, had already put up $2 million for the math department. When Fine, at age seventy, was killed by a speeding motorist while riding his bicycle home one night, the Jones family gave another half million for a building in his memory.

Veblen enthusiastically picked up where Fine had left off. "The principle upon which Fine Hall was designed," he explained, "was to make a place so attractive that people would prefer to work in the rooms provided in this building rather than in their own homes." There were twenty-four "studies" (not "offices"), nine with fireplaces, furnished with overstuffed chairs and sofas, quarter-sawn oak paneling that opened to reveal built-in blackboards and filing cabinets, carved mantelpieces adorned with a heraldic medallion featuring a fly walking along a Möbius strip, and stained-glass windows depicting conic sections and famous mathematical equations. A basement locker room with showers allowed members of the department to "avail themselves of nearby tennis courts" without the need to go home to change. Every time the architects suggested economizing, Jones refused to hear of it. "Nothing is too good for Harry Fine," he would reply.[16]

Veblen, a hopeless Anglophile who had married an Englishwoman, was enthralled by the ancient colleges of Oxford and Cambridge and their rituals, and instituted afternoon tea every day at 4 p.m. in the Fine Hall common room.[17] Princeton undergraduates added a verse to the ever-evolving "Faculty Song":

> Here's to Veblen, Oswald V.,
> Lover of England and her tea;
> He built a country club for math,
> Where you can even take a bath.

Veblen arranged for the nascent Institute to rent space in Fine Hall and share in its tranquil splendors and mathematical collegiality. From the start, too, he saw to it that one of the comforts was astronomical salaries that would free its professors from the need or temptation to undertake outside work to supplement their incomes. Einstein and Veblen received $20,000 a year, Weyl $15,000, von Neumann and Alexander $10,000. Even the lower figures were an unheard of sum for a university professor. The envious members of Princeton's mathematics department coined the enduring wisecrack "Institute for Advanced Salaries" to describe their new, rival counterpart.[18]

Veblen's and Flexner's plan was that younger mathematicians of extraordinary ability, like Gödel, would be invited to come for shorter visits, to have the chance to pursue their research while interacting with its eminent permanent staff. At first the visitors were called simply "workers," a title changed a few years later to "temporary members."[19]

Veblen was not above playing the matter of salaries from both sides of the street when it suited his purposes. When the legal scholar (and future U.S. Supreme Court justice) Felix Frankfurter, a member of the Institute's board of trustees, challenged Veblen's policy of individually negotiating salaries rather than setting them on an objective and equal basis, Veblen loftily explained that the Institute was "a paradise for scholars," who really did not care for money, but only for the search for truth. Frankfurter replied tartly: "The natural history of paradise is none too encouraging as a precedent. . . . Let's try to aim at something human, for we are dealing with humans and not with angels."[20] But if not heaven it was undoubtedly a haven, one that Veblen would work assiduously to expand as a refuge for what was about to become a tidal wave of émigré scientists fleeing Austria and Germany.

LIFE IN PARADISE

The "local beauties of *nature*," Gödel wrote his mother shortly after arriving in the fall of 1933, was Princeton's most striking attribute. As he later described the local landscape to her, "We have a very lovely lake here and around it a very romantic meadow that reminds one of the Prater meadows, even if they are not as large. All in all the country looks much like a park; only the true Alpine forest is missing."[21]

On Nassau Street, the town's tree-lined main thoroughfare, the sparse passing traffic barely disturbed the all-pervading air of genteel calm that enveloped the town—population 12,000, home since 1756 to the famous university of the same name. Settled by Quakers in the sev-

The view down Nassau Street in Princeton in the 1930s,
with the campus to the right

enteenth century, Princeton stood halfway between Philadelphia and New York and near nothing but itself. It had changed remarkably little in the centuries since, still a picturesque pre-Revolutionary village attached to the university campus whose "lazy beauty," as F. Scott Fitzgerald described it in his 1920 novel *This Side of Paradise*, stood at its very center, "the pleasantest country club in America."[22] Nassau Street boasted a traffic light at each end of town, a bank and three churches, a tiny shopping district barely three blocks long anchored by an absurdly quaint half-timbered Tudor-style shop directly opposite the stately gates and neo-Gothic buildings of the university, a brick drugstore that had occupied the same spot since 1858, and the venerable Nassau Tavern, with its laboriously fake colonial taproom, Windsor chairs, low beamed ceiling, massy fireplace and all. The collegiate architecture and enveloping sense of cultural otherworldliness put more than one visitor and academic refugee from the Old World in mind of the great English universities of Oxford and Cambridge, only much more so.

Socially, the atmosphere was another matter. Town and gown equally had a reputation for stodgy conservatism. A jokey letter to the alumni magazine in 1935 called Princeton, "The northernmost outpost of southern culture," but the observation had more than an element of truth: it would be 1948 before the local public schools were integrated.[23] Even while admiring its beauty and peacefulness, Einstein shortly after his arrival summed up the small-town pretensions and complacency that afflicted the self-declared leaders of local "society." Princeton, he wrote to his friend Queen Elisabeth of Belgium, was "a very comically ceremonious, provincial town of lilliputian spindly legged demigods." The word Einstein actually used was *Krähwinkel*—"crow's corner"—a sarcastic German word for a place of petty bourgeois convention, a cultural backwater. He ironically added that he suspected all it would take to secure some welcome privacy was to transgress against the prevailing "fine tone."[24]

The local prudishness was so insufferable to the mathematician Carl Ludwig Siegel that he returned to Germany in 1935 in spite of his strong anti-Nazi and pacifist leanings. (He would make the much more difficult reverse trip a second time in June 1940, fleeing Nazi

Germany for good via Norway and catching one of the last boats to the United States to return to a position at the Institute.) Siegel's unconventional living arrangement, a bachelor sharing a house with two female friends, had aroused the particular censure of the dean's wife. As he disgustedly informed his fellow refugee Richard Courant, "It would be meaningless to escape the sadism of Göring's only to get under the yoke of Mrs. Eisenhart's notion of morality. . . . Please do not be offended that I do not like your America."[25]

To his friend and colleague Georg Kreisel, Gödel acknowledged many years later his own "frustrations of . . . bachelor life in Princeton," attendant on spending a year away from his Viennese girlfriend, Adele. What sexual freedom was possible in cosmopolitan Vienna was definitely not possible in straitlaced Princeton.[26]

Gödel moved into lodgings in a homey Victorian rooming house just two blocks from the campus, at 32 Vandeventer Avenue. Unused to the sleepy ways of Princeton and the early closing hours of its restaurants, he often missed dinner while working late, which worried Mrs. Veblen enough that she sometimes cooked dinner for him herself. As Gödel later observed of the quiet life in Princeton, "If you live in a big city all year, that kind of relaxation in the country can be quite pleasant. With me, it is exactly the other way around. I would love to spend a couple months in a big city to recuperate from the country life here."[27]

Adding to the stuffy social mores of the town and university was a suspicion and resentment of so many foreign scholars, Jews above all. Einstein wrote to a fellow refugee from Nazi Germany hoping to find a place in America, "Carnap told me recently that he had been explicitly told they didn't want to hire any Jews at Princeton. So all that glitters is not gold and who knows how it will be tomorrow. Maybe savages are the better people after all." Those attitudes became fodder for Nazi academic propagandists: snidely noting the Jewish-sounding names in a history of mathematicians in America published a few years later, Wilhelm Blaschke, a pro-Nazi mathematician in Hamburg, wrote, "The most surprising thing is the large-scale mathematical enterprise in the tiny Negrotown of Princeton, where nearly a hundred mathematical lecturers lay their golden eggs almost without students." The word

Blaschke used, *Negerdorf*, was an unpleasant German term for any shithole town, but in the context "Negro" also clearly meant "Jew."[28]

Flexner and Veblen boldly stood up against those attitudes, and not just for humanitarian reasons, but on practical grounds as well. As far as Flexner was concerned, his job was to build up mathematics, period. Writing to the Harvard mathematician George Birkhoff—whom Einstein called "one of the world's greatest academic anti-Semites," and who had given a widely quoted speech warning of the "danger" posed by so many foreign mathematicians invading American universities and taking away jobs from deserving "Americans"—Flexner riposted:

> If we could place fifty Einsteins in America, we would probably within the next few years create a demand from other institutions for several hundred. . . . Hitler has played into our hands and is still doing it like the madman he is. I am sorry for Germany. I am glad for the United States. I will undertake to get a position within a reasonable time for any really first-rate American mathematician, and I will also undertake simultaneously to do the same for any first-rate foreign mathematician whom Hitler may dismiss. The more the merrier.[29]

In more ways than one, the Institute was a haven from the surrounding conservatism of Princeton's town and campus. Von Neumann's unconventional high spirits helped to set the new tone. Israel Halperin, who arrived as a graduate student in the mathematics department the year the Institute opened, recalled the energy that von Neumann's sheer presence gave to life at Fine Hall: "I would come to Fine Hall in the morning and look for von Neumann's huge car, some sort of a convertible, I believe. And when it was there, in front of Palmer Lab, Fine Hall seemed to be lit up. There was something in there that you might run into that was worth the whole day. But if the car wasn't there, then he wasn't there and the building was dull and dead." Von Neumann, recalled his wife Klári, "could not work without some noise or at least the possibility of noise," doing his best work in trains, ships, crowded hotel lobbies. He always kept the door of his office at Fine Hall open,

John von Neumann, with Institute colleagues at tea

and had a running bet with a colleague that if either caught the other working, the guilty party had to pay ten dollars. Von Neumann was never caught.[30]

The large car that Halperin remembered was as much a part of von Neumann's persona as the suit and tie he always wore since once being mistaken for an undergraduate and the parties he and Klári threw at the luxurious home they rented on Library Place, the town's most fashionable address. A notoriously unreliable driver, who according to his daughter had secured his driver's license by the simple expedient of offering the examiner a cigarette from a case with a five-dollar bill discreetly protruding, von Neumann bought a new car at least once a year, whenever he totaled the previous one. A colleague once asked him why he drove such an unprofessorial car as a Cadillac. "Because no one will sell me a tank," he replied.[31]

Having been able to bring his considerable family wealth from Europe, he entertained in lavish style, with cocktail parties featuring uniformed servants bearing copious trays of drinks that offered "an oasis in otherwise somewhat stuffy Princeton" as one colleague said. "The phenomenal feature of von Neumann," recalled Dean Eisenhart's son Churchill, who was studying mathematics at Princeton the year the Institute opened, "was that he could go to these parties and party and drink and whoop it up to the early hours of the morning, and then come in the next morning at 8:30, hold class, and give an absolutely lucid lecture. What happened is that some of the graduate students thought that the way to be like von Neumann was to live like him, and they couldn't do it."[32]

James Alexander, who shared von Neumann's love for hosting alcohol-fueled parties, and who had inherited an even larger fortune— his grandfather had been president of the Equitable Life insurance company—added his own measure of unconventionality to the founding circle of the Institute's faculty with his communist political views and love of mountain climbing. He would always advise incoming graduate students to follow his example and climb in through a second-story window if they needed to get into the library after hours.[33]

None of Gödel's letters to his mother from before the war survive, but he seems to have lived a much quieter life in Princeton than his more spirited colleagues. But mathematically he threw himself fully into life in the New World. To Veblen's suggestion that he hold a seminar or lectures while in Princeton, Gödel replied, "To deliver a course of lectures would be laborious for me in the first months of my stay in America and I would like it better, first to improve my knowledge of English."[34]

Eleven weeks after setting foot in America for the first time he gave what appears to have been his first public lecture in English, at the meeting of the American Mathematical Society and Mathematical Association of America, held on December 29 and 30 in Cambridge, Massachusetts. The title was "The Present Situation in the Foundation of Mathematics," and his handwritten draft of the talk shows the perfect command of the language and great clarity of expression that marked all of his subsequent papers and presentations in English. He began:

> The problem of giving a foundation for mathematics (and by mathematics I mean here the totality of the methods of proof actually used by mathematicians) can be considered as falling into two different parts. At first these methods of proof have to be reduced to a minimum number of axioms and primitive rules of inference, which have to be stated as precisely as possible, and then secondly a justification in some sense or other has to be sought for these axioms, i.e., a theoretical foundation of the fact that they lead to results agreeing with each other and with empirical facts.[35]

In the spring semester, from February to May, he gave a series of lectures in English at the Institute explaining his Incompleteness Theorem, and two of his students transcribed their notes and typed them up for private distribution, with Gödel's approval. On April 18 he traveled to New York to give a popular talk on his theorem to the New York University Philosophical Society, and two days later gave a similar lecture to the Washington Academy of Science with the title, "Can Mathematics Be Proved Consistent?" All of his talks are clear and superbly well organized. But he continued to find public speaking an ordeal, and giving a lecture always left him depressed afterward. Another student who was present for one of Gödel's lectures in Princeton a few years later reported that while the lecture was good, Gödel continued his practice of facing the blackboard the entire time, even though he did not write on it once. "It was clear that he just could not face his audience."[36]

A note at the time in the files of the Rockefeller Foundation, which was undertaking a massive effort to aid scholars fleeing Nazi Germany, noted that Carl Ludwig Siegel had the same problem, but in both cases a place should be found for both men nonetheless: "Both Gödel and Siegel queer—could not teach classes in Univ. Inst. Adv. Study can capitalize their genius."[37] The Institute was only too glad to oblige, and before Gödel departed for home he received an invitation to return the next year. After a leisurely week in New York, Gödel sailed for Genoa on the Italian liner SS *Rex* on May 26 and arrived in Vienna, after a few days spent on the way home in Milan and Venice, a week and a half later. He returned home to a maelstrom.[38]

GOOD REASON TO WORRY

The violent and unsettled scene he returned to had begun with a bloody clash three months earlier. On the morning of February 12, 1934, police in Linz broke down the door of a Social Democratic workers' club to search for weapons. Disobeying a directive from party headquarters in Vienna, the commander of the local socialist Schutzbund militia ordered his men to open fire. Fighting quickly spread throughout the country as Dollfuss's government and the far-right Heimwehr militia saw their chance to crush the Social Democrats once and for all. In the course of three days of clashes, government forces used light artillery to shell the Karl-Marx-Hof, one of the large communal housing projects that were the most prominent emblems of Red Vienna. More than a thousand members of the Schutzbund and other civilians were killed. The following day the Social Democratic party was abolished and its funds confiscated, its leaders who had not fled to Czechoslovakia thrown into prison camps. Nine were executed.

In the months that followed, a series of terrorist outrages by the Nazi Party, which had been banned a year earlier in Austria, were destabilizing the country from the right. Throughout May, June, and July nearly daily bombings and shootings targeted government officials and buildings, power stations and waterworks, and tourism sites important to the Austrian economy. The violence reached its peak on July 25 when 154 members of the Vienna SS, disguised in Austrian Army uniforms, broke into the federal chancellery and shot Dollfuss, mortally wounding him. This time it took the government forces five days to defeat the well-armed insurgents across the country and put an end to the attempted *Putsch*.

Nazi faculty and students in a torchlight march at the
University of Vienna, 1931

The outlawing of the Nazi Party made barely a dent in the giddy
enthusiasm for Germany's Führer among students and faculty at the
University of Vienna. During the 1933–34 academic year, the state
minister of education, Kurt Schuschnigg—who would succeed Dollfuss as premier following his assassination—had had to admonish the
faculty to stop winking at Nazi demonstrations at the university, citing
recent incidents in which students sang the Nazi Party's "Horst Wessel
Lied" and shouted "Heil Hitler!" at the end of lectures.[39]

The crackdown on the socialists was much more effective. Not
long after the government's abolition of the Social Democrats, Moritz
Schlick was summoned to his local police station and informed that
the Ernst Mach Society had been outlawed. "Up with Dollfuss! Down
with the unity of science!" sarcastically responded Otto Neurath. "It
was a tragic spectacle to observe the atrophy of the previously vigorous intellectual life of Vienna," Karl Menger wrote. The Vienna Circle was now routinely "disparaged and maligned" for its "liberal" and

"Jewish" ideas. And in the midst of the summer's turmoil, Hans Hahn, Gödel's mentor, died at age fifty-five following surgery for cancer. The Ministry of Education promptly abolished his chair, a spit in the grave aimed at the dying remnants of the once vibrant Circle.[40]

"After his visit to America," Menger recalled, "Gödel seemed to be rather more withdrawn than before." In June he presented a paper to the Mathematical Colloquium on an extension of his Incompleteness Theorem showing that switching to higher-order systems of logic not only made it possible to prove propositions that are undecidable in a lower-order system, but also often dramatically shortened the length of the proof even for propositions that could be proved in the lower-order system. It was a result that would subsequently have important implications for a vast field of computer science research, the so-called "speed-up" problem of finding ways to shorten computer programs.[41]

At least outwardly, Gödel resumed his accustomed part in the spirited mathematical discussions of the Colloquium, particularly with two new colleagues who would become close friends of later years. During the 1934–35 academic year Alfred Tarski was visiting Vienna on a fellowship. They had met before, during Tarski's first visit to the Vienna Circle and Mathematical Colloquium in February 1930, when Gödel had asked Menger to arrange a private meeting so he could tell the visitor about his recently completed work on first-order logic, but this was the first chance for the two logicians to have an extended discussion of their ideas.[42]

Tarski was brilliant, extremely sensitive about his short stature and Jewish origins, passionately opinionated, "a gregarious person who appreciated and needed the stimulation of other intelligent minds to hone his own competitive edge," in the words of his biographers. Born Alfred Teitelbaum in a wealthy Warsaw family whose fortune, like so many others, had been made in the nineteenth-century textile industry, he bristled throughout his life from the bullies who had taunted him as a "dirty little Jew." At age twenty-two he had converted to Catholicism and adopted his new Polish name, alienating him from his family and forcing him to scramble for jobs to support himself. (Once, when as a young and still unemployed PhD he sought financial help from home,

With Alfred Tarski in Vienna, 1935

his father sarcastically replied, "Money? You need money? Well, why don't you go see your old man Tarski?")[43]

His competitiveness was legendary. To avoid being placed second to Gödel among the world's living logicians, he once described himself to a colleague as "the greatest living *sane* logician." He forever rankled at being beaten by Gödel to the incompleteness result; when he subsequently published an important extension of Gödel's theorem, showing that it is impossible to define the concept of truth within a formal system itself, he insisted to the point of embarrassment that he had obtained most of his results independently and that only "in one place" was his work "connected with the ideas of Gödel." But his affection for Gödel, and later his wife Adele, was genuine and sincere, based on their shared background and zeal for ideas as well as their unspoken competitiveness. Tarski was one of the very few friends whom Gödel ever addressed in his letters using the informal pronoun *Du*.[44]

Gödel's other deepening professional friendship that year was with Abraham Wald. The son of an Orthodox Jewish baker, he had been educated at home because his family refused to allow him to attend the local public schools in Hungary, which required going to class on Saturday, the Jewish sabbath. Taking up mathematics at a late age, he had been one of Menger's most able students, earning his doctorate in three semesters. Menger described him as "small and thin, obviously poor, looking neither young nor old, a strange contrast to the lively beginners." Recognizing what Oskar Morgenstern called Wald's "exceptional gifts and great mathematical power," Menger had approached Morgenstern in his capacity as director of the Austrian Institute for Business Cycle Research to find some funding to support his gifted pupil. "Like everyone else," Morgenstern said, "I was captivated by his great ability, his gentleness, and the extraordinary strength with which he attacked his problems." Wald in turn ended up giving Morgenstern private tutorials in calculus, which led to a warm and enduring friendship.[45]

Wald's research during the year of Gödel's absence from Vienna had focused on price equilibrium in economics, and Gödel, with his usual knack of insight and curiosity, asked Wald to bring him up to date on his work, and immediately offered some useful suggestions on formulating a system of equations to address the question. Wald's subsequent paper—which holds "a paramount position" in the history of economics, in Morgenstern's later appraisal—was a breakthrough not only in explaining the price and quantities of goods produced, but for incorporating the fact that the price of some goods (such as air) is zero when there is an oversupply regardless of how much in demand they are.[46]

Gödel would reconnect with Wald after the war in New York, where he had taken a position at Columbia University after being dismissed by Morgenstern's Nazi successor at the economic institute in Vienna. Wald's death in an airplane crash in India in 1950 was another of the losses of close friends that deeply affected Gödel in the following years.

But outside of mathematical discussions, the emotional withdrawal Menger had noted in Gödel was at times painfully apparent, especially about the darkening political situation. As Menger recalled,

> Gödel kept himself well informed and spoke with me a great deal about politics without showing strong emotional concern about the events. His political statements were always noncommittal and usually ended with the words, "don't you think?" . . .
>
> Even in the face of the seemingly inescapable dilemma between a Europe ruled by Hitler and a second world war, Gödel remained rather impassive. Occasionally, however, he made pointed and original remarks about the situation. Once he said to me: "Hitler's sole difficulty with Austria lies in the fact that he can only appropriate

the country as a whole. Were it possible to proceed piecewise, he would certainly have done so long ago—don't you think?"[47]

Still, Menger worried that Gödel's naiveté might land him in trouble. "In particular I feared, that despite his cautiousness, an injudicious or misunderstood remark of his might have incalculable consequences," he wrote.[48] Menger was worried about something else, too: the signs of deeper instabilities that had begun to surface in his sensitive protégé's demeanor.

"REST CURES"

Gödel's return to Vienna in June 1934 had indeed brought out the first inklings of more serious mental disturbances, in particular the weight loss and insomnia that always were to be an ominous warning of a serious and imminent psychological crisis. "It began," his colleague Georg Kreisel related, "with severe anxiety when he got off the boat."[49] A half year after his return, Gödel wrote apologetically to Veblen to explain that he would not be able to come to Princeton for the spring term as promised, as a result of a bout of "ill health," which he attributed to an infected tooth:

> I had a very bad time this summer. I got an inflammation of the jawbone from a bad tooth and after it was over, I was still feeling wretched for a long time and lost much of weight. In September I had almost recovered but in October I got again temperatures, insomnia, etc. I feel much better now, but am still somewhat sensitive and fear, that a stormy crossing or the sudden change in conditions might bring about a relapse. . . .
>
> I hope that in consideration of what I wrote above you won't be angry with me, that I did not keep my promise of writing you soon.[50]

In fact, in October he had stayed for eight days at a sanatorium just outside of Vienna, in the town of Purkersdorf at the edge of the Vienna Woods. The Sanatorium Westend was a particularly fashionable example of the sanatoriums that catered to the Viennese mania for health cures of all kinds. Designed by the modernist architect Joseph Hoffmann in 1904, its clean simple lines, large windows, and geometric motif of black and white floor tiles repeated in the cubic chairs and furnishings that Hoffmann conceived of as forming a harmonious and restful whole, reflected the pre-Freudian view of nervous disorders as the product of the stresses and unregulated chaos of modern urban life. Borrowing from the tuberculosis sanatoriums of the mid-nineteenth century, which had offered sunlight, fresh air, and healthful food in a tranquil rural setting, the Purkersdorf Sanatorium added a modernist sensibility that was meant to reinforce the therapeutic virtues of remov-

Josef Hoffmann's interior design
at Purkersdorf Sanatorium

Receipt of $50 deposit for Gödel's 1934 stay

ing the patient from unhealthy outside influences and protect him from unpredictable shocks to the nerves. The patient would spend his day in an orderly progression through its serene and purpose-designed spaces: sleeping, eating, physical therapy, and simple leisure activities: there were special rooms for reading, writing letters, billiards and ping-pong, and playing cards.[51]

Health tourism went hand in hand with hypochondria, and "rest cures" were a regular part of Viennese life for maladies of all kinds, real and imagined. Residents at sanatoriums ranged from those suffering from serious mental disorders and tuberculosis, to patients recuperating after surgery, to others simply in search of general rejuvenation. Many advertised themselves as *Sanatorium und Erholungsheim*, the word *Erholung* encompassing a gamut of meanings from "vacation" and "recreation" to "rest" or "convalescence." When Gödel's mother later stayed at Purkersdorf in 1946 while experiencing foot troubles, Gödel wrote to her of his apparently not unpleasant recollections of his own sojourn there while suffering under nervous afflictions: "I still remember very well that park whose trees you write about rustling in their melancholy way. I always liked it very much and it really gave me the impression of a 'castle grounds.' "[52]

Gödel later told Kreisel that during that first episode, the famed Viennese psychiatrist Julius Wagner-Jauregg (who had won the Nobel Prize in Medicine in 1927 for his work on treating syphilis with malaria inoculation) had been called in and found no evidence of "psychosis." Wagner-Jauregg, head of the University Neuropsychiatry Clinic in Vienna, confidently predicted a "recovery in months." In fact, a few weeks later Gödel was present on November 6 for the first meeting of Menger's Colloquium for the fall semester.[53]

But the weight loss Gödel mentioned to Veblen was a foreboding. When he was well he always spoke with pleasure of enjoying the traditional Viennese foods, especially its famous desserts of plum dumplings, *Palatschinken*, or *Guglhupf* with mounds of whipped cream.[54] But his anxieties always first manifested themselves in digestive upsets. In Vienna he had consulted an internist, Dr. Max Schur (who was also an associate of Freud's and later did important work on anxiety in psychopathology), and Dr. Otto Porges, an authority on ulcers and other gastrointestinal disorders. A few years before Gödel's first visit to Princeton, Dr. Porges prescribed for him a long list of forbidden foods (black coffee, strong alcoholic drinks, sharp spices). Gödel's obsessions with stomach upsets are a constant refrain in his letters to his mother, intruding even on memories of otherwise pleasant occasions and outings they shared together.[55]

Aside from worries over his financial situation and the "frustrations of his bachelor life" in Princeton, he was constantly torn by anxieties over revealing himself to others, and his fear he would never be able to live up to the expectations created by his first astonishing successes. "Observation by others is embarrassing to you," he wrote in his 1937–38 diary, "*because you believe they expect something grand*, and by the same token the disappointment is embarrassing just as every deed is embarrassing *because you set your expectations for yourself too high*." Apparently inspired by the Greek Stoics, whose ideas of self-perfection through mental and physical hygiene he encountered in Heinrich Gomperz's philosophy lectures, Gödel filled two entire shorthand notebooks, which he titled "Time Management," with urgent strictures for organizing his life. Noting the paradox that the pursuit of perfection is its own worst enemy—"the main reasons for wasting time and not being able to get many things done is that I want to do things too accurately. . . . If I were to work less, I would do more, above all with more enjoyment"—he created laboriously detailed schedules for balancing work with breaks and diversions, calculating how many minutes to devote at what times of the day to dozens of different activities, analyzed and categorized by type: getting books from the library, talking to colleagues and to family members, going to the bank, getting his hair cut, going for walks, visiting Adele, listening to the radio, ordering pencils and erasers. Lamenting his tendency to delve deeply into every book he picked up, he set himself elaborate instructions for how to read and take notes from a book. Bemoaning the fact that "it takes me five to ten times as long to reach a decision than other people," he filled page after page with procedures to employ in *making* decisions.[56]

Gödel's tragically self-defeating reaction to the fear of never having enough time for his work ("always take on only very little," he repeatedly admonished himself) was to insist that he must do *nothing*, however trivial, without first settling on a clear decision and justification for it, which only sank him further into the abyss of indecision and rigidity. "In general, check every hour what you are doing and whether you are adhering to the program (maybe work using a clock at the desk)," he wrote, and sought refuge in an exacting adherence to formality in his clothes, his correspondence, the neatness of his workspace, as a buttress against the disorder of life. Only Gödel could turn the pursuit of moderation and balance in life into unbalanced obsession.[57]

These traits and others that he later described to Dr. Erlich would likely be diagnosed today as symptoms of obsessive-compulsive personality disorder: extreme perfectionism, a preoccupation with detail to the extent of being unable to finish a task, a compulsion for making lists and notes and following rules, personal rigidity and cautiousness, and an insistence that such behaviors are rational and even desirable. There were also aspects of paranoid personality disorder that were manifest especially during his more serious psychological crises: the fear of being poisoned or drugged against his will, the great reluctance to reveal himself to others, the suspicions that colleagues were against him or seeking to destroy his livelihood.

Yet his sometimes paralyzing indecisiveness and extreme caution about showing imperfections in public were the flip side of an exactness and conscientiousness toward his work that many of his friends recognized as one of his most admirable traits, as both a colleague and a scientist. Years later, Gödel explained why in the 1960s he had turned down an invitation to deliver the prestigious William James Lectures on philosophy at Harvard. It was mostly, he said, because it would be a disservice to the ideas themselves to present them in a less than fully completed form.[58]

Like many others, Hao Wang was struck by how much care Gödel took when asked to review the work of others, even when he fundamentally disagreed with their ideas.[59] Verena Huber-Dyson, a young Swiss mathematician who got to know him later in Princeton, insightfully noted the connection between these two facets of his personality, emphasizing "what a conscientious person Gödel was."

> That goes a long way to explain his reticence, his hesitation to publish or pronounce or do anything that was not honed down to perfect precision and lucidity, and stripped of all superfluous bulk. Less known is how diligently, to the point of pedantry, he scrutinized every proposal, every piece of work or mere argument submitted to him.[60]

In August 1935 Gödel wrote Flexner that he had been in his "normal state of health for some months" and proposed coming to Princeton for the next semester. His second visit proved a complete disaster. He had barely arrived and settled into a rented room at 23 Madison Street, just a block from his previous address in Princeton, when he wrote to Flexner on November 17 that "on account of my very poor state of health I have to resign my position."[61]

Outwardly things seemed to have been going well. On the boat from Europe he had met two other scientists on their way to the Institute for Advanced Study, the theoretical physicist Wolfgang Pauli, who sent Gödel a courteous note asking if he might introduce himself, and Paul Bernays, Hilbert's erstwhile collaborator.[62] Flexner was both sorry and surprised by Gödel's abrupt decision to head home: "I was under the impression that your health had greatly improved, for you certainly looked much better than when you came to Princeton two years ago," he wrote the following day. But Gödel had already decamped on the fifteenth to New York, where he stayed the weekend, returning to Princeton only to pack up for his departure at the end of the month. Flexner gently demurred at Gödel's somewhat wild suggestion that he nonetheless be paid the entire promised stipend of $2,000 for the month and a half he had spent in Princeton, but did agree to send him the balance of half that amount after deducting his advanced travel expenses.[63]

By the time Gödel arrived in Paris on December 7 he was in such a bad state that he phoned his brother to meet him there and bring him home, speaking for over an hour and running up a staggering long-distance telephone charge, equal to nearly $1,000 in today's currency.[64] Veblen meanwhile was so worried that he took it upon himself to cable Gödel's family and let them know the situation. Veblen explained himself to Gödel in a letter which Gödel received only upon his arrival in Vienna:

> December 3, 1935
>
> Dear Gödel:
>
> When I left you on board the *Champlain* in New York I did not intend to interfere any more in your affairs, but after returning to Princeton it seemed that it was not right to fail to let your family know that you were on the way. Otherwise it would be quite possible for some accident to befall you on the way without any of your friends on either side of the Atlantic knowing about it for several days, or perhaps weeks. So I have decided to ask Dr. Flexner to send the following cablegram to your brother: "Returning on account of health your brother arrives Havre December seven via steamer Champlain." I realize that this interferes with your plan of not alarming your family with the idea of your being unwell until you have seen and reassured them. Nevertheless I do not dare to risk the other course of action and I am therefore writing this letter to explain myself and ask your pardon. . . .
>
> My wife joins me in best greetings and the wish to see you again in Princeton in the not too distant future.[65]

A week later, with the protective instinct of so many of Gödel's associates, Veblen sought to buck him up by arranging an invitation for him to give one of the principal addresses to the International Congress of Mathematicians, which was scheduled to meet in Oslo in 1936. During Gödel's short time in Princeton, Veblen told one of the organizers, he had "learned something about his recent, as yet unpublished work," which like his earlier results "is so interesting and important" that it deserved such a prominent forum.[66]

But it was already too late for any such palliative measures. Menger wrote Veblen on December 17 that he had seen Gödel twice since his return and that, as a result of having "overworked himself," he was suffering from insomnia and depression. "It is too bad," Menger reported, that Gödel was having to take sleeping pills to get any rest—"but perhaps still worse" that he had now stopped taking them.[67]

Over the next two weeks his condition rapidly deteriorated. On January 8, 1936, Moritz Schlick wrote to Otto Pötzl, Wagner-Jauregg's successor at the University Neuropsychiatry Clinic and one of Vienna's leading psychiatrists, begging for his help:

Rekawinkel Sanatorium

Please excuse me if I permit myself to recommend my faculty colleague, Privatdozent Dr. Kurt *Gödel*, to your particularly kind attention.

This young man's intellectual abilities are so extraordinary that no praise is too high for them. He is a mathematician and logician and in this field the epitome of a *genius*. Since I am very interested in logic and the logical foundation of mathematics myself, I know his area of work and also know the opinion of his peers regarding him. He is a mathematician of the *very first* order, and his work is universally regarded as *epoch-making*. Einstein seriously described him as the greatest logician since Aristotle, and in fact it is unquestionable that Gödel, despite his youth, is regarded as the foremost authority on foundational issues. Immediately after completing his Habilitation at the University of Vienna, Gödel received an invitation from the famous Flexner Institute in America, where such men as Albert Einstein and Hermann Weyl are working, and last year he received a second invitation from the same Institute, and has now recently returned from America in such deplorable shape.

If Dr. Gödel does not regain his health, it would be a loss of immeasurable consequence for our university and for science throughout the world.[68]

Fearing he might even become violent, Gödel's mother began locking the door to her room at night. Later that month at his family's insistence Gödel entered the sanatorium at Rekawinkel, in the heart of the Vienna Woods, five miles past Purkersdorf down the western railway line. He would be at Rekawinkel and Purkersdorf for four months.[69] In contrast to his lifelong habit of keeping every bill and receipt and doctor's bill and prescription, not a trace of his stay at these sanatoriums in the winter and spring of 1936 is to be found among his papers. Only one hint survives of what might have precipitated the crisis. He told Dr. Erlich four decades later that he was still haunted about one thing that had happened that year: his girlfriend Adele had recently had an abortion, and he could not shake the fear he might be arrested and criminally charged for his part in the affair.[70]

· 7 ·

Fleeing the Reich

ADELE

The first time Oskar Morgenstern met Adele he was appalled.

Morgenstern recorded in his diary his impressions of the woman Gödel had attached himself to. "She is a Viennese washerwoman type," he wrote. "Talkative, uneducated, determined & probably saved his life." He subsequently exhausted the German dictionary for synonyms of "ghastly" or "hideous" to describe her: *schrecklich, fürchterlich, grässlich, abscheulich, nicht erfreulich, eine solche Pest von Frau*— "such a plague of a woman."[1]

Gödel clearly had no illusions about how his educated friends would react to Adele: none of them had met her, or even heard him mention her, until they were engaged to be married in the summer of 1938. "I never met his bride, and only know that 3 years ago when he was ill somebody with the first name Adele visited him," Karl Menger wrote Oswald Veblen on hearing news of his engagement.[2]

In saying that Adele "probably saved his life," however, Morgenstern was speaking the literal truth. More than once her resolve and decisiveness rescued Gödel from catastrophe. The first time was during his terrible crisis in the spring of 1936 when, deep in the throes of paranoid delusion, he refused to eat, convinced that the doctors in the sanatorium were secretly putting drugs in his food or trying to poi-

Adele in 1932

son him. Adele patiently sat with him hour by hour, coaxing him to eat by first tasting each spoonful of food that she personally prepared and brought to him every day.[3]

In a portrait from the time when they first met she has a certain charm and youthful good looks, with short, stylishly waved hair revealing a graceful neck and shoulders. She often wore high heels to try to add to her short five-foot stature; her hair was blond, her eyes grey. The left side of her face was marred by a large port-wine birthmark that she was nearly always careful to keep turned away from the camera when her photograph was taken, or which she covered with varying degrees of success in a heavy layer of face powder.

But by the late 1930s the seven-year age difference was already beginning to tell, exaggerated by Gödel's perpetual skinniness and her increasingly zaftig hausfrau physique. Together, they often looked more like a middle-aged matron and her little boy than a romantic pair. There were more than a few hints of her dominating role in their relationship. "A mystery. . . . He must be completely under her thrall," Morgenstern wrote in his diary, using the word *hörig*, which literally means "bondage," but which often carries the connotation of sexual power or control.[4] In some of his notebook passages discussing sex, Gödel expresses a clinical view of "coitus" as a simple matter of mental and physical hygiene, on a par with getting a good night's sleep or regular exercise. A "visit to Adele," he observed, has a wonderful way of clearing his mind— "a walk has a similar effect," he added. But elsewhere he acknowledged being drawn increasingly to do her sexual bidding. "In order to evoke the feeling that 'life has meaning,' it is very useful to satisfy her in some better way (possibly perversely)," he wrote. "Idiocies with Adele," he tersely noted in another entry: "Sadism and the like (no pure love)."[5]

His guilt over her pregnancy and abortion, along with her genuine devotion to him during his ordeal in the sanatoriums, were in any case a powerful claim on his loyalty to her. But a number of their Austrian and other German-speaking friends who got to know Adele later in Princeton saw a different side to her and to their relationship, and even Morgenstern softened his initially harsh judgment with a more nuanced view of her complex personality. "She can also be very nice; that she means well always is clear," he wrote in his diary—adding that she did not have an easy time of it either, with a husband like him.[6]

As Morgenstern acknowledged, she was also a superb cook, with a repertoire of classic Viennese dishes—wiener schnitzel, dumplings with sauerkraut, whole carp, cakes, plum dumplings, vanilla crescents—that delighted their Austrian friends. Still, Morgenstern could not refrain from observing after one evening with the Gödels, "Good food, bad entertainment because Frau Gödel talked too much." He frequently complained it was impossible even to have a conversation with his friend when Adele was present.[7]

Nonetheless, Georg Kreisel, Alfred Tarski, and Verena Huber-Dyson all admired Adele's high spirits, wisecracking observations about Princeton life, and bantering style with her husband, which he seemed to enjoy. "It was a revelation to see him relax in her company," Kreisel recalled. "She had little formal education, but a real flair for the *mot juste*." She affectionately called him "Kurtele," or *strammer Bursche*, strapping lad; she made fun of his reading about ghosts, telling him that *he* sounded like an old Viennese washerwoman with his fascination on that topic; and she habitually referred to the Institute for Advanced Study as the *Altersversorgungsheim*, the home for old-age pensioners—which she depicted in humorous invented accounts as teeming with pretty girl students lined up at the doors of the eminent professors. (Kreisel was able to divert the conversation to a different topic the time Adele derided him for being an old Viennese washerwoman by observing that actual Viennese washerwomen probably did not rely on books for their information.)[8]

Occasionally driving back from a party from which Kurt had stayed home, Adele would "heave a deep sigh and say what a relief it was to be with such good friends without having to worry about a genius in the car," Huber-Dyson related. "At that time I felt a vague kinship with her, both of us balancing, a bit precariously, at the edge of the society we were living in."[9]

Tarski enjoyed a particularly warm relationship with Adele that was more than just a manifestation of his laying-on of Old World charm. Deeply touched by a cake she sent him when he was living in Berkeley, California, and homesick for Vienna and his old friends, he wrote back at once thanking her for her gift, which he told her "affects my heart at least as strongly as my sense of taste. Please do not forget me." He sent her stamped envelopes and stationery so she would have no excuse not to write back to him. "I would very much like to see you again, to chat with Adele, and to discuss various questions with Kurt," he once wrote to them.[10]

In the darkest moments during his confinement in the sanatorium in the first months of 1936, convinced he was going to die, Gödel was tormented by the thought that the important work he had told Veblen about the previous fall might never be completed. He had begun work on two important but controversial axioms of set theory, the Axiom of Choice and the Continuum Hypothesis, and had already worked out a proof showing that both were consistent with the other fundamental axioms. John von Neumann later told Morgenstern how, during one of his frequent trips to Europe at this time, he had visited Gödel in the sanatorium just as he was falling apart, and the extraordinary scene that had unfolded:

> Johnny related that Gödel some years ago, just as he became unhinged, detailed his entire proof about the Continuum Hypothesis to him, under a seal of secrecy. So clearly, that Johnny could have published it, in Gödel's name, in case G died. G had also possibly communicated it to Menger. At the same time he knew what was happening to him. Then he got better, & he finished it here in P[rinceton] in 1938/9. What strange things go on in the human brain.[11]

Gödel was later convinced that the treatments he received at Purkersdorf and Rekawinkel had done him permanent harm. He remained haunted by a persistent fear of being secretly drugged, which in his worst moments burst forth in full-blown hallucinations. During his most serious episodes of paranoid delusions in the 1970s, he claimed that at both sanatoriums he had been given a heart stimulant, strophanthin, in his food and by injections in the night, and that that was now happening again: someone was unbolting his door and sneaking into his room at night and injecting him with the drug against his will.

It is not impossible that he received even more drastic treatment than that when he was at Purkersdorf. Dr. Manfred Sakel, developer of the horrific and almost surely useless treatment optimistically termed insulin shock "therapy," was practicing at Purkersdorf at the time. A precursor to electroshock treatments, Sakel's procedure involved repeatedly placing a patient in an insulin-induced coma. The unintended side effects included extreme restlessness, major convulsions, the possibility of permanent brain damage, and death. Although there is no evidence Gödel was ever subjected to such an ordeal, he always claimed that the "strophanthin" that he believed he had been given there had a permanent effect on his brain. Adele told him that after he left the sanatorium that spring of 1936, he never seemed "quite as intelligent" ever again.[12]

"MY MURDERER"

No sooner did Gödel return to Vienna from Purkersdorf than a violent incident struck the University of Vienna, which deeply shook his remaining reserves of security.

For six years, Moritz Schlick had been stalked and harassed by a former student. Johann "Hans" Nelböck, who came from a small village in rural Upper Austria, received his doctorate under Schlick in 1931.[13] He was a mediocre student, earning the lowest possible passing grade for the thesis he submitted, but was filled with ideas of his own philosophical genius and obsessive resentment of the professor who failed to appreciate his gifts.

A few years before completing his studies he had become equally obsessed with a fellow student, Sylvia Borowicka, the daughter of a wealthy Viennese family who was completing a thesis under Schlick as well. Her topic was "The Agreeable and the Beautiful in Philosophy."

Borowicka had no romantic interest in her megalomaniacal pursuer, but for whatever reasons chose to let him know that she did fancy her professor, and thought she detected signs that the interest was reciprocated. Crazed with jealousy, Nelböck began phoning Schlick at home at night, loitering outside his house, following him onto trams, finally telling Borowicka that he planned to shoot his professor, and then himself, with a pistol that Borowicka had for reasons unclear given him. By the summer of 1931 Borowicka was sufficiently alarmed to inform Schlick, who informed the police, who arrested both Nelböck and Borowicka, charging them with illegal possession of a firearm.

Both were referred to Dr. Pötzl of the University Neuropsychiatry Clinic for evaluation. Pötzl found Borowicka to be "a nervous girl of slightly eccentric character," but Nelböck, he concluded, was a full-blown schizoid psychotic. After being treated for a few weeks at a mental institution, Nelböck was released to the care of his family.

Within a few weeks he was back in Vienna, now studying for a teaching certificate. Schlick and his family were spending the 1931–32 academic year in Berkeley, California, where he was a visiting professor, but as soon as he returned in May 1932 the harassing calls started up again. Nelböck began confronting him at the university, shouting that he had engaged in "immoral games" with Borowicka, vowing again to shoot Schlick and then himself. In the summer of 1932 he was committed to the psychiatric ward of Vienna's General Hospital, ruining his chances of ever obtaining a teaching position once the university authorities were informed. He was now even more firmly convinced that Schlick was to blame for everything that had gone wrong in his life.

Ignoring a court order barring him from returning to Vienna, he was soon back at the university yet again tormenting his mild-mannered and increasingly fearful professor. Schlick never doubted that Nelböck was in earnest: he began referring to him as "my murderer." Karl Menger, in his memoir, described a scene of haunting foreboding that occurred in the spring of 1936. Menger had received an invitation to a formal opening of an exhibition at the Hofburg where the president of the Republic was to officiate. "I arrived a little late," Menger recalled, "and found that the room was so packed with guests that it was almost impossible to move."

The first familiar person whom I saw was Schlick standing not far from the entrance. We were talking when a passage opened in the crowd near us. Through it, the President, having completed the opening ceremony, was leaving with his party. To my surprise, one of the men behind the President waved rather intimately to Schlick. . . .

"You have friends in the government?" I asked teasingly. But Schlick's expression changed to one of utmost seriousness and he said in a grave tone, "That is not a friend. That is a security man who used to be my body-guard. . . . For some time now I have been threatened by an insane person who is in and out of mental institutions; and the man behind the president used to be assigned to my protection."

"So you are no longer threatened," I said.

Schlick sighed. "Until quite recently the fellow had been interned," he said. "But just three days ago he has been released again and yesterday I had another threatening telephone call from him. Yet, for all his threats, he has never actually harmed me. So I don't dare complain to the police again." And, as though it had happened only yesterday, I remember how Schlick added with a forced smile, "I fear they begin to think it is I who is mad."[14]

A few weeks later, on June 22, 1936, Nelböck left his apartment at eight o'clock heading for the university, where Schlick was to give a lecture that morning. In his pocket was a .25-caliber pistol, loaded with seven bullets. Nelböck waited in the main university building on the first steps of the grand staircase leading to the Faculty of Law, knowing that Schlick would ascend the opposite staircase to the Faculty of Philosophy. At 9:20 Schlick appeared and began slowly walking up the staircase. Nelböck ran up the stairs past him, turned, and fired four times at point-blank range, putting two bullets through his heart, one through his bowels, and one into his lower leg. "Now, you damn bastard, there you have it!" he shouted. Schlick's lifeless body was carried into the Philosophical Faculty meeting room. Nelböck, having appar-

Der Attentäter Dr Hans Nelböck

Das Opfer des Mordanschlags Professor Dr Moritz Schlick

Das erste Verhör mit dem Mörder am Schauplatz der Tat.

Moritz Schlick's murder depicted in a Viennese tabloid

ently forgotten his vow to kill himself in the excitement, stood calmly, pistol in hand, waiting for the police to arrive and take him away.

Speaking freely to the examining magistrate, he made no attempt to deny his act, but now had a new story to justify his vengeance on his former professor. It was his duty, he said, to oppose the atheistic and destructive ideas of positivism, through which Schlick had insidiously tried to destroy the deeply held religious beliefs that he had grown up with. At his trial, Nelböck declared to the judge, "For me, Schlick's behavior expressed the utter lack of scruples of his so-called scientific worldview." As the judge summarized Nelböck's testimony at his trial, his former professor had "robbed him of his love, of his creed, and of his means of livelihood."

Sentenced to ten years in prison, Nelböck immediately became a

Hans Nelböck at his trial

martyr of the anti-Semitic and anti-liberal right, particularly at the university. An anonymous article by "Prof. Dr. Austriacus" which appeared on July 12 simultaneously in two widely read far-right Austrian Catholic journals, *Die schönere Zukunft* and *Das neue Reich* ("The More Beautiful Future" and "The New Reich," the founder of the former having declared his mission that of combating the "bad press" of "Jewish hacks"), expressed perfunctory disapproval of the murder before making it perfectly clear who the real villain in the affair was. The notorious leader of the "Vienna Circle" was not a true philosopher, Dr. Austriacus observed, "but only a physicist." He had denied not only the existence of God, the human soul, and moral law, but metaphysics itself—thus dismissing the "supreme discipline of the humanities" as nothing but a "triviality." It was therefore no surprise that Schlick had gathered around him "all the Jews and freemasons" who shared this contempt for Christian morality.[15]

A typescript copy of the article was circulated among Schlick's colleagues and friends, Gödel among them, who kept his copy among his papers for the rest of his life.[16] Dr. Austriacus, affecting an air of scholarly analysis, suggested that all the newspaper articles reporting on the "sensation" of the crime had "failed to address the true facts and motives behind this terrible case." Nelböck had been driven to his desperate act "under the influence of the radically destructive philosophy which Prof. Schlick had been teaching at the university since 1922; that is to say this bullet was not guided by the logic of some lunatic looking for a victim, but rather by the logic of a soul deprived of its meaning of life. . . . I know of several cases myself where young students have lost all faith in God, the world, and humanity under the influence of Schlick's philosophy."

Dr. Austriacus was none other than Johann Sauter, a professor on the Faculty of Law and an illegal member of the Nazi Party, who among his other notable views had once denounced Freud's theories as "simply an elaborate form of pornography disguised as science." Schlick's murder, he explained, had similarly laid bare "the disastrous influence of Judaism":

Now, the Jewish circles will certainly not tire of praising him as the greatest of thinkers. We understand that very well. For the Jew is the born anti-metaphysician. In philosophy, he loves logicism, mathematicism, formalism, and positivism—those very ideas which Schlick united so completely in himself. We would, however, remind everyone that we are Christians living in a Christian–German state, and that it is *we* who will decide which philosophy is good and suitable. The Jews can have their Jewish philosophers in their own cultural institutes! But in the chairs of philosophy at the University of Vienna in Christian–German Austria, there belong Christian philosophers! It has been recently explained repeatedly that a peaceful solution to the Jewish question in Austria is in the interests of the Jews themselves, since otherwise a violent solution is inevitable. Hopefully, the terrible murder at the University of

Vienna will serve to accelerate a truly satisfactory solution to the Jewish question!

Within months of the Nazi takeover of Austria in March 1938, Nelböck was paroled on a petition from Sauter and others, who portrayed him as a victim of the *Systemzeit*—the Nazis' odd, disparaging term for the pre-Anschluss period in Austria—and Schlick as "an exponent of Jewry."[17]

For the dwindling remnant of the Vienna Circle, Schlick's violent end was in many ways the final blow. Ernest Nagel, an American philosopher who had visited Vienna the year before during a lengthy tour assessing the state of analytical philosophy in Europe, wrote in early 1936 that his impression was that the Circle had already "seen its best days": "Rudolf Carnap's departure for Prague, Otto Neurath's exile to The Hague, Kurt Goedel's visits to Princeton, Hans Hahn's premature death, have deprived the Circle of some of its most original and strongest members." Within weeks of Schlick's murder Menger received a cable offering him a visiting professorship at Notre Dame University in South Bend, Indiana, which he accepted at once, thereby spelling the effective end of the Mathematical Colloquium as well.[18]

Gödel, his brother later recalled, was deeply shaken by Schlick's murder. Since his release from Purkersdorf he had continued to return to the sanatorium for monthly examinations. But in October 1936 he left Vienna for a retreat to more peaceful surroundings, this time accompanied by Adele. They registered at a hotel in Aflenz, in the Styrian Alps, as "Herr und Frau Dr. Kurt Gödel," and there on a high Alpine meadow Adele continued to nurse him back to health, bringing his weight back up from 100 to 140 pounds, still having to eat spoon for spoon with him from his plate because he feared he was being poisoned.[19]

THE YEAR OF LIVING INDECISIVELY

"I found it very hard to give up the Colloquium," Menger later wrote, "but in Vienna the time for meetings of the type of ours was obviously running out, while I hoped that I might have the opportunity to develop a similar group in America. And it was very sad to leave Gödel, Wald, and the others, but I trusted that before long I should see them in the free world." Interviewed in 1978, Menger said it had been obvious to him that a catastrophe for Austria was coming. "I could not understand that many people did not see that. It was as clear to me as anything. I of course did not suspect the details, but that it must lead to catastrophe, was to me wholly clear."[20]

Deeply worried about Gödel, he wrote from his new home in the free world to Franz Alt, asking him to keep their friend out of trouble:

I am deeply saddened to be able to do so little for the splendid circle of Viennese mathematicians so dear to me. I believe you should all get together from time to time, and especially should see to it that Gödel takes part in the colloquium. It would be of the greatest benefit not only to all the other participants but also to Gödel himself, though he may not realize it. Heaven knows what he might become entangled in if he does not talk to you and his other friends in Vienna now and then. If necessary, be pushy, on my say-so.[21]

In the summer of 1937, after a very successful first semester at Notre Dame which led to an offer of appointment at the university, Menger approached its president, Father John Francis O'Hara, and told him about Gödel and his work. O'Hara, "an extraordinarily energetic and resourceful priest," as Menger described him, was itching to build a first-rate mathematics program at Notre Dame, and to that end was quite open to inviting "scholars of many countries to his school." O'Hara immediately asked if Gödel might come for a semester as a visiting professor.[22]

Everyone among Gödel's circle in Vienna was by now desperately looking for a situation, anywhere but home. Within a year Franz Alt would leave for the Econometrics Institute in New York City; Abraham Wald for the Cowles Commission for Research in Economics in Colorado Springs, Colorado; Friedrich Waismann for Cambridge University; Oskar Morgenstern for Princeton University. Karl Popper had already departed on a fellowship to Cambridge, then to a position at Canterbury College in Christchurch, New Zealand—"halfway to the moon," he said. Marcel Natkin was permanently settled in Paris, having abandoned philosophy altogether for a career as an artistic and portrait photographer. Olga Taussky had spent a year at Bryn Mawr, the women's college in Pennsylvania, and then two years on a fellowship at Cambridge where she spent most of her final year desperately applying for jobs that would allow her to remain in England. In the fall of 1937 she was a lecturer at one of the women's colleges at the University of London, where she had agreed to take on an inhuman course load—nine classes a week, with homework and exams to be graded in all—as a condition of the job.[23]

For much of that fall of 1937 Gödel filled his diary with records of the conversations he had with his friends in which he asked them at length about their experiences abroad and the salaries and conditions of various possible fellowships or other positions, in England and

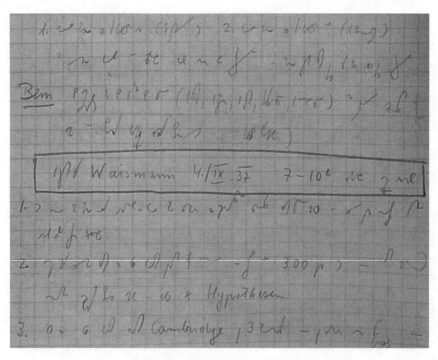

Gödel's 1937–38 shorthand diary

America. Taussky, in Vienna for a visit in September, had a long talk with him, but aside from offering some amusing accounts of the drafty English buildings ("in England one could run a windmill with windows closed—poorly heated and damp, therefore also everyone in England is *schmattisch* and stunted," she said, using the Yiddish word for bedraggled or worn-out)—she had little to offer in the way of solid leads on jobs. She did bring some news of Wittgenstein, who had also just finished a fellowship at Cambridge, where he made himself "very unpopular" with his monumental impatience: "If one enters his room, it is at the risk of being thrown out." She also reported that Wittgenstein was said to have discussed "with four people the philosophical implications" of Gödel's work.[24]

But Gödel himself was paralyzed by indecision about his future. A serious complication was that the family's money was now starting to run low. He was still doing nothing to try to generate a regular income, even though he had not much more than the equivalent of a few thousand dollars of his own left in bank accounts and bonds from his father's legacy, plus what he had saved out of his stipends from the Institute for Advanced Study. When, at the end of the year, Edgar Zilsel attempted to organize a biweekly meeting to keep alive a remnant of the old Gomperz and Schlick circles and one person after another objected to each proposed day because of other engagements, Gödel ruefully observed, "I am the only one who has no conflict on any day."[25]

For her part, Gödel's mother, finding life in Vienna ever more expensive and with her savings locked up in Czechoslovakia, decided to move back to the villa in Brno, which meant that if Gödel were indeed to stay in Vienna he would need to find an apartment of his own by the end of the year.[26] He exhaustively interviewed friends and his uncle Carl Gödel about rent control laws and where to look for a suitable place, all the time making elaborate calculations of how much he might earn, and save, if he were to go to America again. He wrote out one formula, subtracting expenses (including the $2,000 he had run up staying at Rekawinkel and Aflenz), and adding back what he would save by not having to rent an apartment in Vienna:

salary – living Am. – travel – Rekawinkel – Aflenz +
savings + saving exp. Vienna ≃ D

Ship   Paris   train   Italy   insurance
3900 – 1000 – (800 + 200 + 200 + 100 + 100) –
2000 + 400 + 400 ≃ 300$[27]

Menger had written to Gödel over the summer promptly relaying
President O'Hara's offer and enclosing information about Notre Dame,
but still Gödel dithered. He had written back to Menger expressing
interest, but was hesitant to commit himself:

I would be agreeable in principle to coming to the University of
Notre Dame; indeed it would interest me very much to get to know
the workings of a Catholic American university. The bulletins
which were sent to me interested me very much and I would like to
express my thanks for them. . . . For my part, the summer semester of 1938 would be the earliest that could be considered. . . . It
would be essential to me to be under obligation for *one* semester
only. As you well know, I have had bad experiences with my health
in America, and would therefore not like to be bound in advance
for a longer period. But if the above mentioned requirements are
met, I would gladly accept, assuming the remaining conditions
(salary and duties on my part) are acceptable.

There is not much new with me. Since I returned from Aflenz
things have been worse with me again health-wise, though still tol-
erably well. . . . At the moment I am considering whether I should
lecture next semester on something introductory or something
higher-level, or whether I should not lecture at all and use the time
for my own work. In the second case, there is the danger that I
won't have any students at all, as not enough preparation in formal
logic or set theory is available now that Carnap and Hahn are no
longer lecturing.

In closing, I would like to congratulate you warmly on your
appointment at Notre Dame University, of which I learned from

the bulletin, even though I also very much regret that I am thereby again losing a friend in Vienna.[28]

In July Gödel had also put out a feeler to von Neumann during his last swing through Vienna about returning to the Institute for Advanced Study. Von Neumann warned him that at the Institute "they are tired of foundations" of mathematics, but advised him strongly "not to lose any time but to accept everything in America immediately." It was not just Jewish scientists who were under attack by the Nazis. Von Neumann mentioned a recent article in a Nazi weekly, *Das Schwarze Korps*, that vilified the German physicist Werner Heisenberg as a "white Jew" for his work in quantum mechanics, which placed him squarely at the center of the "Jewish conspiracy" in physics. The official Nazi line—promoted by among others the Nobel Prize winners Johannes Stark and Philipp Lenard—not only endorsed Hitler's crackpot ideas about "cosmic ice" but propounded an entire racial theory of science, in which *Deutsche Physik*, based upon exact observation, represented the superior creation of "the Nordic-Germanic blood component of the Aryan peoples," while such "mathematical fabulations" as Einstein's relativity theory and quantum mechanics were the insidious products of "the Jewish spirit," with its degenerate wallowing in "ego" and "self-interest." Gödel's work on foundations of mathematics was dangerously tainted with that same "Jewish spirit" of abstraction.[29]

On November 1 Gödel received a cautious reply from Veblen, who suggested that the best he could offer was funding for a brief visit to the Institute to give a few lectures: only $1500 was available.[30] Meanwhile, despite von Neumann's counsel and Menger's urging, Gödel had decided to turn down the offer from President O'Hara. The letter went astray, but Menger finally cabled for an answer, and in December Gödel brought him up to date on his latest indecision, as well as his ongoing work on the Continuum Hypothesis:

I have decided not to come to America after all in the current academic year, as I also already definitively wrote in the letter to President O'Hara which was lost. At the moment I cannot even commit for Notre Dame for 1938–39, but I will be able to write you something definite in about two months. . . .

Last summer I continued my work on the continuum problem and I finally succeeded in proving the consistency of the Continuum Hypothesis with respect to general set theory. But I ask you for the time being to say nothing about this to anyone. Besides you, I have so far informed only von Neumann, to whom I sketched the proof during his last stay in Vienna. Right now I am also trying to prove the independence of the Continuum Hypothesis, but do not know yet whether I will pull it off.[31]

What all of his backing and forthing really came down was his uncertainty on what to do about Adele, who was eager to move in with him permanently while he was plagued by recurring doubts, annoyed by her spending and tears and interruptions of his work, and wondering if it was within his power not to marry her if it came to that.[32]

On a list of pros and cons of going to America, Gödel included under the minuses "miss out applying in Vienna for math position"— Taussky had relayed to him Menger's suggestion that "it would be lovely" if Gödel were to get his position at the University of Vienna if Menger decided to remain permanently at Notre Dame—while under the pluses, besides "pleasant sea travel, Venice, New York, music, Aflenz, Paris, useful English," he added, "Problem Ad[ele] temporarily taken care of."[33]

Austrian marriage law, which under a 1934 concordat with the Vatican gave the church complete legal jurisdiction over marriages by its members, barred civil divorce for Catholics. Although Adele was long separated from her husband, she and Gödel could never marry under the current situation.[34] But that November, at last biting the bullet, Gödel rented an apartment in the far-off suburb of Grinzing, high in the hills overlooking Vienna, and Adele moved in with him. He no doubt chose the remote location to keep his living arrangement out of scrutiny of those who knew him. Adele set about ordering repairs and

Himmelstraße in Grinzing, 1938

improvements—a double steel sink in the kitchen, brass and chrome ceiling lights, the bills addressed to Herrn & Frau Dr. Gödel.[35]

Grinzing was a world apart from the city below, a quaint, almost Alpine Austrian village at the edge of the Vienna Woods, its sweeping hillsides above covered with the vineyards that supplied the local white wine. From the Schottentor tram stop two blocks from the university, just opposite the Ephrussi palace on the Ringstraße, the ride on the 38 tram took twenty minutes, past the Votiv Church on the left, the Votiv Kino movie theater, and Sigmund Freud's office at the bottom of the steep hill on Berggasse to the right, then on past the Café Josephinum which Gödel had lived above in 1929, then the long climb into the green hills with a fleeting view of a distant castle above before arriving at the loop through the covered archway of the Grinzing station at the end of the line. A ten minute walk up Himmelstraße—"Heaven Street"—took him to his new flat, in an elegant and solidly built apartment house on practically the last street within the Vienna city limits.

Zilsel's efforts to organize a last hurrah of the philosophical and mathematical discussion groups of earlier years bore some fruit; Gödel himself gave a paper to the circle in January 1938 on the possibilities of keeping alive Hilbert's program in a revised form, with a series of purely constructivist consistency proofs for various parts of mathematics. Gödel kept a record of the papers and discussions of the group: there were talks on dialectical materialism and Marxist theory, religion and science, some unpublished papers by Schlick on the existence of meaningful propositions; Gödel was keenly interested in the studies of personality types that one member of the circle, the young Jewish psychologist Else Frenkel, was carrying out under the direction of the famous psychologists Charlotte and Karl Bühler at the university's Psychological Institute.[36]

But it was a far cry from the old circle of Menger, Schlick, Hahn, Carnap, Taussky, and the other leading lights of bygone days. Zilsel constantly annoyed him with his "threatening" questions and teasing insistence on addressing him as "Professor." Rose Rand, a longtime hanger-on of the Schlick Circle, who was living off practically nothing while suffering delusions of philosophical grandeur—von Neumann had described her to Gödel that summer as a megalomaniac who had reproached him for not inviting her to America ("Waismann and Gödel, they're small lads in comparison to her," von Neumann wisecracked to Gödel)—put her head on her knees and slept through the entire discussion.[37]

On February 21, 1938, Flexner wrote to Gödel inviting him once again, with "the greatest pleasure," back to Princeton the following year, for the first or second term as he preferred, with a stipend of $2,500 and the freedom to go to Notre Dame whenever he chose.[38]

Three weeks later hell descended on Vienna.

MARCH VIOLETS

In what would later be recalled as a strange omen, the Northern Lights appeared across Austria that winter, blazing forth in an unusual display. In one last bold gamble to save Austria from Hitler's campaign of economic, diplomatic, and military pressure aimed at fomenting a Nazi takeover, Chancellor Schuschnigg on March 9 announced a snap plebiscite to take place just four days later, on Sunday, March 13, to affirm Austria's independence.

American press correspondents in Vienna estimated that the Nazis would gain 60 percent of the vote on a referendum asking Austrians if they wanted to join Hitler's Germany. Schuschnigg's referendum accordingly was worded in a way to appeal to Austrian patriotism, Christianity, and German nationalism all rolled into one: a simple yes or no on the proposition, "For a free and German, independent and social, Christian and united Austria, for freedom and work and for the equality of all those who declare for race and Fatherland." A massive government propaganda campaign swung into action, covering walls and sidewalks with slogans urging citizens to vote "Ja," bombarding towns with leaflets dropped from aircraft, and broadcasting nonstop messages of support for the Fatherland over the airwaves.[39]

The next night a strong south wind arose, and all the next day, under a cloudless sky, the city was blanketed with plebiscite leaflets whipped up in the violent gusts. From the north came not vague omens but more tangible threats. Responding with rage, Hitler massed troops on the border and delivered Schuschnigg an ultimatum: abandon the referendum, or resign and allow the Austrian Nazi Party to take charge of the government. Schuschnigg prudently did both. The next morning columns of German troops streamed into the country. The operation was so hastily planned that trucks ran out of fuel and had to stop at Austrian gas stations to refill their tanks, but otherwise the invaders encountered no obstacles and no resistance. That afternoon hundreds of Luftwaffe bombers appeared over Vienna, literally darkening the sky in a show of unmistakable force.[40]

By evening Hitler was in his hometown of Linz, addressing a rapturous throng assembled before the town hall. That night in Vienna, wrote the German playwright Carl Zuckmayer, "hell broke loose."

> The underworld opened its gates and vomited forth the lowest, filthiest most horrible demons it contained. The city was transformed into a nightmare painting by Hieronymus Bosch; phantoms and devils seemed to have crawled out of sewers and swamps. The air was filled with an incessant screeching, horrible, piercing, hysterical cries from the throats of men and women who continued screaming day and night. Peoples' faces vanished, were replaced by contorted masks: some of fear, some of cunning, some of wild, hate-filled triumph.

Zuckmayer had seen a dozen battles in the Great War, barrages, gas attacks, going over the top; he had been a bystander during Hitler's 1923 Munich *Putsch*; he had lived through the first years of Nazi rule in Berlin. "But none of this was comparable to those days in Vienna," he wrote. "What was unleashed upon Vienna had nothing to do with the seizure of power in Germany, which proceeded under the guise of legality. . . . What was unleashed upon Vienna was a torrent of envy, jealousy, bitterness, blind, malignant craving for revenge."[41]

Zuckmayer barely escaped a mob that had surrounded his taxicab, brazening it out by shouting a gruff "Hei'itler" in his best imitation of a Prussian Army sergeant and giving the Hitler salute out the window. Jews were beaten and humiliated in the streets, the beards of Orthodox Jewish men cut off by grinning Hitler Youth and jackbooted SA men, synagogues ransacked and Torah scrolls torn to shreds, Jewish children forced to paint "Jude" on the windows of Jewish-owned establishments. Throughout the city Jewish shops were looted, not even with violence but with simple impudence, as customers helped themselves to the merchandise and walked out without paying.[42]

It was, however, less avarice than pure sadism that reigned in the days immediately following the Anschluss. The Jewish essayist and coffeehouse intellectual Alfred Polgar paid sarcastic tribute to his countrymen's zeal. "The Germans are first-class Nazis, but lousy anti-Semites," he observed. "The Austrians are lousy Nazis, but by God what first-class anti-Semites they are!" That first night, a mob broke into the Palais Ephrussi, swarmed up its grand staircase and smashed their way through the house, finally dragging a large ornate desk to the hallway and pitching it over the handrail onto the flagstones of the courtyard below. "We'll be back," they told the terrified family as they left.[43]

Walter Rudin, then a high school student, later a distinguished professor of mathematics at the University of Wisconsin, was another eyewitness to those nightmarish days and nights. "One of the popular local entertainments the first couple of weeks was to round up some Jews—preferably elderly, it was more humiliating for them—force them to their knees, hand them a small brush and a bucket of lye (to burn the skin on their hands) and make them scrub political graffiti off the sidewalk, much to the delight of the onlookers." The graffiti were the myriad slogans urging *Ja* on the referendum of the now instantly expunged era of Austrian patriotism. Rudin's gym teacher, whom he had always hated, showed up at school the next day strutting in a storm trooper's uniform complete with pistol strapped to his belt. A few days later Rudin heard that he had accidentally shot himself in the foot, "one of the very few bits of cheerful news at the time."[44]

Cardinal Innitzer, who as rector of the university in 1929 had clamped down on rioting by German-nationalist students, now ordered the city's church bells rung and swastika banners displayed. Buoyed by his delirious reception, Hitler decided to annex Austria outright. On March 15, in a speech before a madly cheering crowd of a quarter of a million who jammed into the Heldenplatz, the great square before the Hofburg, he declared, "In this hour I can report to the German people the greatest accomplishment of my life. As Führer and Chancellor of the German nation and the Reich, I can announce before history the entry of my homeland into the German Reich." Austria ceased to exist. Its territory within the Reich was now merely *Ostmark*, "The Eastern Marches," a name dating to its ancient identity as the borderlands of the German Holy Roman Empire.

The first days of *wilde* Aryanization of Jewish property by ungoverned mobs were soon succeeded by the far more efficient machinery of the Nazi bureaucracy that arrived in force to take charge. "Commissioners" were appointed to take over Jewish businesses, and Adolf Eichmann was dispatched from Berlin to begin the process of systematically stripping Jews of their assets as the price of a now desperately coveted exit visa. "In one respect we were better off than the German Jews," Rudin recalled. "There the screws were tightened gradually, and for the first couple of years there was hope that it would all blow over, that a different government would be formed, that things would get back to normal. As a result, many German Jews procrastinated until it was too late. In Austria it became absolutely clear within a couple of days that the only option was to get out." Sigmund Freud, after much international pressure, was permitted to depart, even to take some of his furniture and prized archaeological collections with him, but in return was forced to sign a dictated statement declaring "that I have been treated by the German authorities, and especially by the Gestapo, with the esteem and regard due to my scientific reputation; that I have been able to pursue my activities freely; and that I do not have the least reason for complaint." Freud told his son Martin that he was tempted to add a postscript in the style of the testimonials in commercial advertisements: "I can very much recommend the Gestapo to anyone."[45]

At the University of Vienna, a *kommissarische Rektor* was appointed and immediately launched a "purification" (*Säuberung*) of the faculty and student body of all Jews. Two thousand students were dismissed, along with 97 of the 297 professors and dozents in the Faculty of Philosophy and 180 of 315 in the Faculty of Medicine. Menger, still officially on leave from his permanent appointment at the University of Vienna, took the small satisfaction of beating the Nazis to the punch and cabled his resignation before he could be fired. All licenses to teach for dozents were also temporarily suspended by order of the government on April 22, as Gödel was informed by letter the next day from the new *kommissarische* dean, Viktor Christian. Even those not dismissed outright were required to reapply as a "Dozent of the New Order" if they wished to resume their duties. Gödel was thus deprived even of his nominal employment.[46] (As an academic authority on "The

Nazi takeover at the University of Vienna

Jewish Question" and an officer in the SS, Christian would during the war amass a priceless library of rare Judaica plundered from the Nazis' victims, as well as a collection of "anthropological material" consisting of skulls and skeletons dug up from Vienna's Jewish cemetery at Währing.)[47]

So many Austrians rushed to take out memberships in the Nazi Party—the "March violets," they were sarcastically called—that the authorities stopped accepting new applications. Adele dutifully sent in hers, along with the 2 Reichsmark fee, but was never enrolled. Her mother, Hildegarde Porkert, however made it in time (membership No. 2,654,956, issued April 2, 1938), while her father, Josef Porkert, had actually been a member since 1932, boasting the much lower number 1,451,013.[48]

Gödel would for years be wracked with guilt that he—like 99.74 percent of Austrians, if the results of a second plebiscite that Hitler called for April 10 are to be believed—had voted with the Nazis to approve the incorporation of Austria into the Reich. He did so, he explained,

Receipt for Adele's Nazi Party membership application

because he desperately needed a passport to get out of the country, and feared the consequences of defying the new regime.[49]

AROUND THE WORLD IN 274 DAYS

On July 6, 1938, the Nazi government issued a new marriage law that brought Austria into conformity with existing German law, which placed marriage and divorce completely under civil auspices.[50] That meant Adele was free to marry. In another hint of Gödel's passive acquiescence to Adele's determined hold on him, at the end of August he signed a power of attorney authorizing her to make all of the arrangements for their wedding. On September 20, two weeks before he left for Princeton, they were married in a small civil ceremony, with nine friends and family in attendance, among them Adele's parents, Gödel's cousin Carl Gödel, and his brother Rudi, who was meeting Adele for the first time. Gödel preserved the receipt for their frugal reception, held in the Rathauskeller in the basement of the city hall: 11 cups of bouillon and 11 appetizers, 27 pieces of pastry, 2 glasses of sherry and 4 of sparkling wine, 2 bottles of mineral water.[51]

"I, too, think that marriage may be quite good for him," Menger wrote to Veblen. To their surprise, Gödel arrived in Princeton on October 15 by himself. "Why didn't you let your wife come? That would surely be much nicer for you," Menger wrote his friend later that fall.[52] But that may have been putting things the wrong way around: a more likely inter-

Wedding portrait, September 1938

pretation is that he had agreed to marry her as a condition for her agreeing to let him spend the year away from Vienna, as he had been planning to do before their marriage was even a possibility.

In any case he seemed in far better health than on his previous visits to America. His work on the Continuum Hypothesis was progressing well, and he expressed himself amenable to giving lectures at the Institute and teaching a regular class or seminar at Notre Dame, where he had agreed to spend the spring semester. Although hesitant to offer an introductory course in mathematical logic, as Menger had proposed, because of his "inadequate English," he graciously acquiesced after a lengthy correspondence discussing how he could best contribute to Menger's program at Notre Dame.[53]

Though appreciative of Gödel's "conscientiousness and his desire to be of use," Menger was realistic enough to see that his dream of transporting his Mathematical Colloquium from Vienna to the American heartland was going to take more than just a visit from the world's greatest living logician to make happen. Gödel's class began with about twenty students; half of them were young doctoral candidates and instructors in mathematics who stuck with it to the end, the other half older philosophers who dropped out almost immediately, uncomfortable with the entire approach to formal logic that was at odds with Catholic traditions of logic. Menger recalled,

> During his stay at Notre Dame, Gödel appeared to be in fairly good health, but not particularly happy. He lived on campus . . . but he had quarrels with the prefect of his building for various trivial reasons (because of keys and the like). I found it not always easy

to settle the differences, since the prefect was an old priest, set in his ways, while Gödel was emphasizing his rights.[54]

Gödel's insistence on his rights exasperated Menger even more when he tried to apply such logical reasoning to the situation back in Vienna, regarding what he considered the Nazis' "illegal" action suspending his dozentship. "How can one speak of rights in the present situation," Menger asked him in frustration. "And what practical value can even *rights* at the University of Vienna have for you under such circumstances?" Toward the end of the semester, Menger recalled, Gödel, "who until then had been his usual dispassionate self, appeared to be restless." He spoke frequently of wanting to return home to Vienna at the end of the semester to be with his family. "Despite pleas and warning by all his acquaintances at Notre Dame and Princeton, he was determined to go," Menger wrote, "and he went."[55]

Besides the obvious consideration that he had left Adele behind, the fact was that he could not legally stay in America past mid-July. On his earlier visits he had obtained an immigrant visa that permitted multiple reentry to the United States. But the Germans had canceled all Austrian passports following the Anschluss, and with the crush of desperate refugees seeking asylum from Hitler, he was able to enter the country only on a nine-month visitor's visa when he arrived for his current stay on October 15, 1938. To extend his stay he would have to leave the country and apply for a new visa from abroad in any case.

Two days before Hitler's invasion of Poland launched the Second World War, Menger received a note from Gödel that, as he drily put it, "may well represent a record for unconcern on the threshold of world-shaking events." "Since the end of June I have been here in Vienna again and had a lot of running about so it was unfortunately not possible to write up anything for the Colloquium," Gödel wrote. "How did the examinations turn out for my logic lectures? . . . In the fall I hope to be back in Princeton." Menger's subsequent coolness toward his old friend puzzled Oskar Morgenstern, who thought Menger must be jealous of Gödel or have some strange paranoia about visiting Princeton. But as Menger later acknowledged, Gödel's political naiveté and heedlessness of the greater suffering of so many others made it "not easy to find in me all the warmth that I used to feel for him." He did not find it as easy as Gödel's other friends to forgive the timidity and small-mindedness that sometimes went hand in hand with conscientiousness and winning innocence.[56]

Gödel returned home to a bureaucratic nightmare. He could not leave again without permission of the Nazi government; worse, the authorities kept delaying month after month his required army medical examination, and would do nothing while that was pending.[57] The Institute had offered him a position at the much increased salary of $4,000 a year for 1939–40. But with the outbreak of the war, the United States had stopped issuing visitor's visas, and the quota for immigration visas from Germany and Austria was so oversubscribed by desperate Jews seeking to flee the country that that avenue was now barred, too.

Von Neumann, throwing himself into the problem with his usual energy and steely intellect, refused to concede defeat. "Gödel is absolutely irreplaceable," he insisted to Flexner. "He is the only mathematician alive about whom I would dare to make this statement. Salvaging him from the wreck of Europe is one of the great single contributions anyone could make." A week later, on October 5, he took it upon himself to cable Gödel in Vienna, "INVITATION STILL VALID COME AS SOON AS POSSIBLE."[58]

He followed up by drafting for Flexner a masterful legal analysis of the situation, having discovered a route that would allow the U.S. consulate in Vienna to issue Gödel a visa at once. Certain professions, "professors" among them, were eligible for a special "non-quota" immigration visa, not subject to country-by-country quotas. The catch was that to be defined as a professor, the applicant had to show he had "followed the vocation of professor continuously for at least two calendar years immediately prior to applying for admission." But von Neumann found a loophole in that, too: if the applicant's vocation had "been interrupted for reasons beyond his control," he could still qualify under certain circumstances. Von Neumann's memorandum went on to present a thoroughly persuasive if slightly mendacious argument that that was indeed the case for Gödel.[59]

But getting a U.S. visa was only half the problem. Without assurance that Gödel would return, the German authorities would not grant him permission to leave—and so he rightly feared that having a U.S. immigration visa would ring alarm bells in the Nazi bureaucracy. Meanwhile, hedging his bets, he formally applied to become a Dozent of the New Order. Unlike the old dozentships, the new positions carried a regular salary. The Nazi official in charge of university affairs at the Ministry of the Interior and Cultural Affairs, Friedrich Plattner—also newly appointed as a professor of physiology at the university, as well as being a high-ranking SS staff officer—had already proposed to the rector that the simplest way to deal with dozents like Gödel "in whom further lecture activity is unwelcome" was simply to stall their applications past the October 1 deadline, when all of the old dozentships will "definitively expire."[60]

But Gödel managed to submit his application nonetheless, and the head of the National Socialist Dozents Association was asked to report on his ideological bona fides. The heaviest mark against him was his work on foundations of mathematics, which like Einstein's Theory of Relativity was now scorned as "Jewish science":

The former Dozent Dr. Kurt Gödel is scientifically well considered. His Habilitation was carried out under the Jewish Professor Hahn. He is accused of always having moved in Jewish-liberal circles. It must be mentioned here though that in the *Systemzeit*, mathematics was heavily Jewified. Direct comments or activity against National Socialism have not been made known to me. His professional colleagues have not got to be closely acquainted with him, so further information about him has not been obtained. It is from this not possible for me to expressly support his appointment as Dozent of the New Order, but neither do I have the grounds to speak against it.

Heil Hitler!

Dr. A. Marchet, Dbdf.

Dozentenbundsführer

d. Universität Wien[61]

Gödel had temporized in another significant way, closing his bank account in Princeton on September 20 and moving more than $1,000 to Vienna, which required him to file an official explanation for the source of that foreign currency—which he signed with a "Heil Hitler!" for the one and only time in his life.[62]

In late November, to make matters still worse, he finally had his medical exam and was improbably declared fit for garrison duty, liable to be called up for military service at any time. A story filtered back to Menger that, attracted by his spectacles and intellectual appearance, a group of Nazi thugs had beaten him on the street, taking him for a Jew—until Adele drove them off by brandishing her handbag or umbrella. (Years later Gödel sarcastically remarked to Kreisel that only Austrian Nazis were capable of such *Schlamperei* in misidentifying him.)[63]

Meanwhile, the new director of the Institute for Advanced Study, Frank Aydelotte, a former president of Swarthmore College, bombarded the German embassy and the U.S. State Department with appeals trying to break the impasse. No doubt swallowing hard, he obsequiously wrote the German chargé d'affaires in Washington on December 1, making the good Nazi case for permitting Gödel's return to Princeton:

Dr. Gödel is an Aryan; he is thirty-three years old and married. He would, I suppose, be liable for military service, but since he is one of the greatest mathematicians in the world, we very much hope that the German government will in this case think it more important for him to continue his scientific work. . . . There is no question but that his presence here makes a significant addition to the prestige of German science.[64]

The wheels of the bureaucracy had in fact begun to turn. The dean of the University of Vienna was asked about the advisability of granting Gödel's requested leave of absence, and replied:

Gödel enjoys particular repute . . . in his area of work, which encompasses the border area of mathematics and logic that Gödel's teacher, the Jewish Professor Hahn, particularly favored; Gödel is especially well esteemed in the USA where these questions of the foundations of mathematics are of wider interest.

Regarding the political views of Gödel, I have drawn upon the counsel of the University Dozentenbundsführer Prof. Marchet, whose opinion coincides completely with my own personal impression. Gödel, who grew up at the time when Vienna's mathematical community was completely under Jewish influence, possesses hardly any inner relation to National Socialism. He gives the impression of a completely apolitical person. Therefore he will hardly be able to cope with the difficult situations that, for a representative of the New Germany in the USA, are sure to arise.

Personally, Gödel makes a good impression; in this regard I have never heard a complaint against him. He has good manners, and certainly will commit no social mistakes that will harm the reputation of his homeland abroad.

If Gödel should be denied travel to America for political reasons, the question of his livelihood arises. Gödel has no income whatsoever here and only wants to accept the invitation to the USA in order to be able to support himself. The whole question of his leaving would be moot if it were possible to offer Gödel a correspondingly paid position within the Reich.[65]

On January 2, 1940, Gödel wired the triumphant news that he and his wife had received their exit visas. Two weeks later he cabled from Berlin that he had his American "professors visa" in hand and they were departing at once. Because German passengers crossing the Atlantic on Italian or other neutral ships were liable to arrest by British and French warships, the German authorities required him to go the long way around, by way of Russia and Japan. He spent his final days in Berlin shuttling from one embassy to another filling their passports with the required transit visas, before he and Adele boarded the train that would carry them on the first leg of their journey, through Nazi-

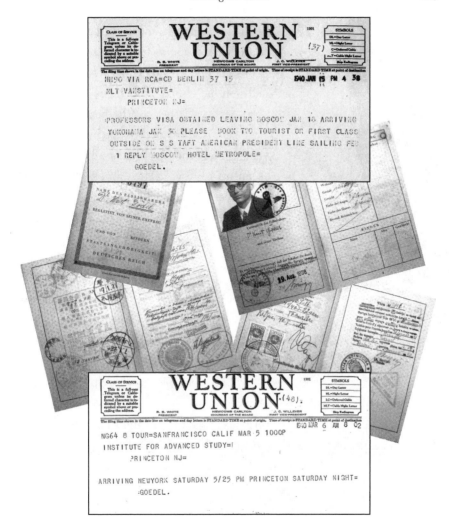

From Vienna to Princeton, 1940

occupied Poland, Lithuania, and Latvia, to Moscow, expecting at every moment to be arrested or sent back.[66]

They stayed overnight in Moscow at the Hotel Metropole, then boarded the Trans-Siberian railroad for the six-thousand-mile journey across the bleak emptiness of the Russian winter to Vladivostok. Then by steamer to Yokohama, where they waited two weeks for the first American ship, the SS *President Cleveland*, while Adele filled her time

On shipboard across the Pacific

and an entire suitcase purchasing Japanese knickknacks to decorate their new home.[67] The *President Cleveland* called at Honolulu, which Gödel described to his brother in a letter as "the best part of the journey," before docking at San Francisco, "absolutely the loveliest of all cities I have seen up to now." Then across the American West by train to New York, arriving in Princeton on the evening of Saturday, March 9, nearly two months after the start of their incredible adventure. In nine months Gödel had traveled completely around the world. He would never leave the East Coast of the United States again.[68]

Morgenstern saw him the next day, and asked him how things were in Vienna.

"The coffee is terrible," Gödel deadpanned.

"He is very amusing," Morgenstern wrote in his diary that night, "with his mixture of depth and unworldliness."[69]

· 8 ·

New Worlds

CONTINUUM AND CHANGE

"Gödel was well prepared to like America," said his friend Georg Kreisel.[1]

Three weeks after his arrival, he wrote to his brother in Vienna with a greenhorn's wonder at the cheap items for sale on the shelves of the local Woolworth's store in Princeton, where he and Adele went searching for bargains as they set up housekeeping in a furnished apartment at 245 Nassau Street. "There are the so-called 10 cent-stores," he enthusiastically told Rudi, "where all household articles are almost given away. e.g., a nice glass sugar bowl costs 5 cents and a little framed picture (very charming) the same. (How long the things last only time will reveal!)"[2]

To improve his English he soon took up the habit, which he continued for years, of meticulously recording unfamiliar slang or technical words he encountered reading the Sunday *New York Times*—*jaywalker, frigate, chaff, Kangaroo Court, humdinger, barnstorming*—along with facts he thought he should know to better fit into American society: the titles of the leading popular magazines; the proper amounts to tip taxi drivers, doormen, barbers; how public housing in New York City and athletic scholarships at colleges work; the names of people and organizations in the news.[3]

His relief at his deliverance was tempered by worries over family and friends he had left behind. Back in Brünn—since March 1939 a part of the Nazi-occupied Protectorate of Bohemia and Moravia—his mother was finding things even harder, with inflation, wartime shortages, and a precipitous drop in her income. "That Redlich once again reduced Mama's pension by half I find unbelievable," Gödel wrote to Rudi. "Do you really think the factory is doing so poorly?" He and Adele sent off packages of food—especially hard-to-come-by items like coffee, tea, and cocoa—not sure if they would get through or be confiscated or pilfered, and he offered to contribute 25 Reichsmark a month, about ten dollars, as soon as he could to their mother's living expenses. However, "I am also currently not exactly swimming in money," he explained. "Only half of the stipend for 1939/40 was paid out and of that more than half went for the trip. I even had to borrow money, of which I still owe 50$. On 1 July payment of the stipend for the next year is to start; that will rapidly improve the situation."[4]

The last remaining few members of the Vienna Circle had finally managed to escape the Reich in the months following the Anschluss: Rose Rand and Edgar Zilsel from Vienna, Philipp Frank from Prague. But Otto Neurath, who had taken refuge in Holland years before to avoid arrest during the Dollfuss regime, was caught off-guard by the Nazi attack on France and the Low Countries in May 1940. Gödel anxiously wrote to Veblen later that summer to ask if he had any news of his old colleague. As he would later learn, Neurath's escape was even more dramatic than his own: after scrambling aboard a highjacked fishing boat in Schveningen harbor as Nazi bombs rained down, he had been rescued from the foundering boat by a passing British destroyer, only to be interned as an enemy alien on his arrival in England. From a prison camp on the Isle of Man, Neurath wrote with aplomb to a friend. "I have always been interested by the conditions in British jails," he insisted, "and would gladly have paid for this information—but now I can get it for free!" Through Bertrand Russell's and Albert Einstein's intercession, he was released at the end of the year, and offered a lectureship at Oxford University.[5]

By the time of Gödel's return to Princeton, the Institute had moved

The Institute's newly completed Fuld Hall

into a home of its own, on 265 acres of former farmland and woods, a mile to the west of the main university campus and separated from it by a golf course. Gödel later described to his mother his new serene surroundings: "Everywhere there are trees in bloom and the area around Princeton is like a large park. The Institute is somewhat on the outskirts and I take a lovely daily walk of ca. ½ hour there in the midst of blooming trees and bushes behind which houses with their large front yards are hidden." The major parcel of land on which the new campus sat, the Olden Farm, had been the site of a victorious but bloody battle by George Washington against British troops in the Revolutionary War. Gödel had an office on the second floor of the new building, Fuld Hall, looking out to the rear, with "a very nice view onto a huge meadow with the edge of the woods on the horizon."[6]

Flexner from the start had vociferously opposed spending money on a building, arguing that every dollar the Institute spent buying land or constructing its own offices was a dollar unavailable for salaries. In 1937 the mathematicians were still sharing space in Fine Hall, the economists were crowded into a rented basement in the Princeton Inn, and Flexner himself had an office in a commercial building on Nassau Street opposite the university campus and didn't see why everyone shouldn't be content to go on that way indefinitely. "I would far rather rent additional floor space," he told Veblen, "and get our minds so full of the purposes for which we exist that we will all become relatively indifferent to buildings and grounds." He thought he detected in Veblen's contrary view the atavistic avarice for land of a Norwegian farmer's son. "He is a most excellent person," Flexner archly commented, "but the word 'building' or 'farm' has an intoxicating effect upon him." But Veblen prevailed, the founders ponied up another half million dollars, and Veblen got to reprise his earlier role in the construction of Fine Hall by working out every last detail in the design of Princeton's newest country club for math. "The prospect of a visit from the architect usually costs Professor Veblen a day's work and a night's sleep," Flexner reported.[7]

Fuld Hall was a Georgian red brick edifice, a deliberate contrast to Princeton's university Gothic, standing alone in the midst of a vast sweeping lawn, fronted by a circular drive. Except for the eerie absence of visible life it looked like any traditionally designed university or prep school main hall; as it was, it had the feel more of a sanatorium, orphanage, or old age home. On the top floor was a dining room that served lunch and dinner at steeply subsidized prices, and a few steps down from the ground floor was a common room with French doors leading out to the lawn and comfortable chairs and sofas, where Veblen promptly reinstituted his daily afternoon teas.

Gödel began his new life there with a burst of renewed professional energy. At the Institute he again gave lectures, even if, as usual, he lost most of his audience by the end of the class; Yale University invited him to present a talk on intuitionist logic; and most of all he was hard at work putting into final form his complete proof of the independence of the Continuum Hypothesis, which he already had published two short notes about in the *Proceedings of the National Academy of Sciences*. George Brown, a student who had taken detailed notes of Gödel's 1938 lectures at the Institute on the subject, produced a draft and was helping him prepare the manuscript for publication by Princeton University Press.[8]

Next to his Incompleteness Theorem, it was to be his most significant mathematical contribution, one that went to the heart of the foundations of set theory, the concept of infinity, and Gödel's own philosophical ideas about mathematical truth and reality. The Continuum Hypothesis had first been proposed by Georg Cantor in 1878, and had been a source of mathematical controversy ever since. In the course of his pathbreaking work on infinite sets, Cantor made the remarkable discovery that infinity comes in different sizes. The smallest infinity is the kind represented by the ordinary counting numbers, 1, 2, 3, . . . As Galileo had observed three hundred years earlier, even infinite sets of objects that at first glance might appear to be larger or smaller than the infinite set of counting numbers usually turn out on closer inspection to be exactly the same size, since they can be listed side by side with the counting numbers, in what mathematicians call a one-to-one correspondence. Thus, there are as many even numbers or squares of numbers as there are counting numbers, as both of those sets can themselves be lined up in 1, 2 , 3, order—they are "countable":

2	4	6	8	10	12	14	16	. . .
$\updownarrow$	$\updownarrow$	$\updownarrow$	$\updownarrow$	$\updownarrow$	$\updownarrow$	$\updownarrow$	$\updownarrow$	
1	2	3	4	5	6	7	8	. . .

1	4	9	16	25	36	49	64	. . .
$\updownarrow$	$\updownarrow$	$\updownarrow$	$\updownarrow$	$\updownarrow$	$\updownarrow$	$\updownarrow$	$\updownarrow$	
1	2	3	4	5	6	7	8	. . .

Even the fractions constitute an infinite set no larger than the counting numbers, since with a clever enough scheme one can place all of them in a 1, 2, 3, progression as well—for example by lining them up in sequence according to the sum of their numerator and denominator:

1/1	1/2	2/1	1/3	3/1	1/4	2/3	3/2	4/1	1/5	5/1	. . .
$\updownarrow$	$\updownarrow$	$\updownarrow$	$\updownarrow$	$\updownarrow$	$\updownarrow$	$\updownarrow$	$\updownarrow$	$\updownarrow$	$\updownarrow$	$\updownarrow$	
1	2	3	4	5	6	7	8	9	10	11	. . .

But Cantor was able to show that, by contrast, the set of all points on a line, the so-called "continuum"—which is precisely equivalent to the set of all real numbers, which includes "irrational" numbers like pi and the square root of 2 that can be expressed only as an infinite series of decimals—are *not* countable by any possible scheme: they constitute a larger infinity than that of the counting numbers.

Cantor's argument was ingenious and simple. Imagine any list of real numbers between 0 and 1; then create a new number by selecting one digit in turn along a diagonal line running through them:

$$0 . \mathbf{3} 4 9 0 1 1 7 \ldots$$
$$0 . 1 \mathbf{5} 8 0 2 2 8 \ldots$$
$$0 . 9 6 \mathbf{7} 1 4 0 5 \ldots$$
$$0 . 2 3 1 \mathbf{4} 1 5 9 \ldots$$
$$0 . 7 7 4 1 \mathbf{0} 6 3 \ldots$$
$$0 . 8 3 1 1 9 \mathbf{7} 5 \ldots$$

If you then alter that new number (0.357407. . .in this case) by adding one to each of its digits, the resulting number (0.468518 . . .), Cantor observed, will *never* appear on the list, however infinitely long it is. It cannot be the first number on the list, since its first digit, 4, differs from the first number's first digit; similarly, it can't be the second number on the list, since its second digit differs from the second number's second digit; in fact it can't be the nth number on the list whatever n may be, since its nth digit will always differ from the nth number's nth digit. Whatever way you try to list the real numbers in a countable fashion, it is always possible to generate another real number not on the list.

Cantor called the size of an infinite set its "cardinality," and used the Hebrew letter aleph to designate the different sizes of infinity: $\aleph_0$ (pronounced "aleph null") is the cardinality of the counting numbers; successively large infinities are designated $\aleph_1$, $\aleph_2$, and so on. The cardinality of the continuum is accordingly one of those alephs greater than $\aleph_0$.

Of particular importance to set theory is the fact that the number of all subsets that can be formed from the infinite set of counting numbers is also one of those larger alephs—in fact, it is exactly equal to the cardinality of the continuum. The general rule is that a set of n elements will have 2^n possible subsets. From a set with two elements, a and b, four ($= 2^2$) different subsets can be formed, starting with the "null set" { } which contains no elements: { }, {a}, {b}, {a, b}; likewise from a set with three elements, eight ($= 2^3$) subsets can be formed: { }, {a}, {b}, {c}, {a, b}, {b, c}, {a, c}, {a, b, c}. So by the same rule, the number of subsets that can be formed from an infinite set with cardinality $\aleph_0$ will be $2^{\aleph_0}$, which will thus equal one of those larger alephs.

A crucial question that much of subsequent set theory depends upon is whether any infinite subset of the real numbers can have a cardinality that is smaller than that of the continuum, yet larger than that of the counting numbers. In other words, should the cardinality of the continuum be designated as $\aleph_1$, meaning it is the very next infinity after the counting numbers (in which case $2^{\aleph_0}$ equals $\aleph_1$)? Or should it be designated by some larger aleph, such as $\aleph_2$, meaning there exists some infinity between the counting numbers and the continuum (and $2^{\aleph_0}$ is thus greater than $\aleph_1$)?

Cantor's Continuum Hypothesis proposed that the cardinality of the continuum is indeed $\aleph_1$: in other words, every infinite subset of the real numbers is going to be the same size as either the set of all real numbers or the set of counting numbers; there is no infinity in between the two. The great importance of the question for mathematics was reflected in Hilbert's placing the question of the cardinal number of the continuum at the very top of his famous list of unsolved problems in 1900.

Gödel's proof took a major step toward answering the question—marking the third time he had succeeded in tackling one of Hilbert's famous challenges—by showing that Cantor's Continuum Hypothesis is at least *consistent* with the other basic axioms of set theory; that is, its contradiction cannot be derived from them. He proved the same for the Axiom of Choice, the other fundamental proposition of set theory that remained on strangely ambiguous footing. One can accordingly add either or both of these propositions to the axioms of a consistent system of set theory and not have to worry about generating any new contradictions in the results derived from them.

But that did not say anything about whether the Continuum Hypothesis is true or not, and Gödel remained firmly convinced that it should be possible to establish its absolute truth or falsity. In an article entitled "What Is Cantor's Continuum Problem?" which he was subsequently asked to contribute to the *American Mathematical Monthly*—part of a series of "What Is . . .?" articles in the magazine aimed at a nonspecialist audience—Gödel strongly leaned toward the belief that it in fact was *not* true. "It is very suspicious," Gödel wrote, "that, as against the numerous plausible propositions which imply the negation of the continuum hypothesis, not one plausible proposition is known which would imply the continuum hypothesis. Therefore one may on good reason suspect that the role of the continuum problem in set theory will be this, that it will finally lead to the discovery of new axioms which will make it possible to disprove Cantor's conjecture."[9]

The next step, to which Gödel would come tantalizingly close over the next two decades, was to prove the independence of the Continuum Hypothesis from the other axioms—that is, to establish that it cannot be derived from them. If the Continuum Hypothesis is both consistent with the axioms of set theory (thus not *disprovable*) and independent of them (thus not *provable*), that would affirm the inadequacy of existing theory, as well as offering a stunning instance of the kind of undecidable proposition that, as his Incompleteness Theorem demonstrated, will exist within any consistent mathematical system.

FRIENDSHIP AND DISTRUST

Georg Kreisel, like his other close friends, thought that Gödel's tendency to keep to himself did not stem from of any fundamental antisocial inclination, but rather from wariness at what it might cost him in time and energy. "He is commonly taken to be a hermit," his friend the logician and philosopher Hao Wang said, "however he was very warm-hearted to personal friends." Wang, who first met Gödel in 1949 and corresponded with him for years on technical matters before Gödel finally agreed in 1971 to meet with him regularly to discuss his philosophical views, was one of the privileged few to break through that wariness. "An important point is: he limited his obligations in every possible way," observed Wang. "Only a few people had close contact with him, but *if* anyone connected closely with him, it was very closely."[10]

Kreisel once asked Gödel why, since both he and Adele so obviously enjoyed being hospitable and having friends to visit, they did not have people over more often. Gödel replied that he "had noticed that most people showed more excitement in company than they felt, and he found this very tiring." ("Clearly," observed Kreisel, "at times he needed very few data to reach, painlessly, a very sound conclusion.")[11]

His devotion to his friends was evidenced in his refusal to join the general disapprobation of Kreisel in the Princeton and larger mathematical communities when, in 1957, Kreisel ran off with Verena Huber-Dyson, who was then married to the English physicist Freeman Dyson. Before they left, Kreisel brought Verena and her two small children to tea with the Gödels one afternoon, possibly indirectly seeking Gödel's approval of the affair. Kreisel subsequently paid homage to "Gödel as a staunch friend—when I disturbed some people in the '50s by what were then considered to be indiscretions."

That was an understatement. Kreisel and Huber-Dyson subsequently appeared together at mathematical conferences and social gatherings where he introduced her as "my wife, Mrs. Dyson."[12]

But Gödel kept up with Kreisel devotedly, with long transcontinental phone calls when he and Verena moved to California, and he used his expense fund from the Institute to have Kreisel come regularly to Princeton to keep him up to date with developments in mathematical foundations. Huber-Dyson, having been bored to tears by Princeton and motherhood, later said that "being with Kreisel was not fun, but it was

Verena Huber-Dyson, 1954 self-portrait

meaningful," reawakening her life as a woman and a mathematician with much yet to contribute. Not just a brilliant logician but a ravishing beauty, she subsequently had a long on-again off-again affair with Alfred Tarksi, whose abusiveness did not make up for *his* brilliance.[13]

There was by now a large community of German and Austrian exiles living in Princeton. The literary scholar Erich Kahler and his wife, Alice, welcomed a constant stream of refugees, both as dinner guests and as temporary lodgers, to their home at 1 Evelyn Place—among them the writer–mathematician Hermann Broch, who lived with them from 1942 to 1948. Paul Oppenheim, a chemist and philosopher of science who became a close friend and neighbor of Einstein's, hosted a regular lunch for émigré German scientists, artists, and intellectuals at his home at 57 Princeton Avenue, just around the corner from Einstein's house, right after his weekly Sunday morning walk with Einstein. Gödel had contact with both circles, but never very closely.

His two great friends from the time of his return to Princeton, Albert Einstein and Oskar Morgenstern, were in many ways the unlikeliest. He had met Einstein at least formally on his earlier visits, but it was only from 1940 on that their friendship blossomed. Paul Oppenheim later claimed credit for their more intimate acquaintance. Aware of Gödel's shyness in approaching his famous colleague, even though their offices on the second floor of Fuld Hall were only a dozen feet apart, Oppenheim marched into the hallway, knocked on both doors simultaneously, and when the two men emerged, announced, "Einstein, this is Gödel, Gödel, this is Einstein." Oppenheim ever after described this as his "only contribution to science."[14]

Einstein had followed his own advice in earnest about purchasing privacy for himself by violating the stuffy social mores of Princeton. He regularly went around in an old leather jacket, baggy trousers, shoes with no socks, jokingly referring to himself as a "museum piece" and telling friends, "At Princeton they regard me as the village idiot."[15] He lived in a modest, white-painted cottage-style house near the Institute on 112 Mercer Street, which—only in Princeton—would subsequently be home to two other Nobel Prize winners.

But he could not escape fame. Once on his daily walk to the office a car passed him, then proceeded to run off the road directly into a tree as the driver turned to gawk at the instantly recognizable figure of the world's most renowned scientist. Einstein remarked to Gödel and Morgenstern that the autograph seekers who all too often hounded him "are the last remnants of cannibals; in both cases they want to acquire the spirit of the devoured."[16]

Within a few years of their reintroduction in 1940, Gödel proudly confided to his mother that he had been to Einstein's house several times for scientific discussions: "I think in general it only happens rarely that someone is invited to his house," he reported. Their friendship was subsequently cemented in the deep and wide-ranging daily conversations they enjoyed while walking to or home from the Institute. "Gödel . . . was the only one of our colleagues who walked and talked with Einstein on equal terms," said Freeman Dyson. Ernst Straus, Einstein's mathematical assistant from 1944 to 1948, considered Gödel to

With Einstein

be "certainly by far Einstein's best friend" of those last years of his life. "They were very, very dissimilar people," Straus observed, "but for some reason they understood each other well and appreciated each other enormously":

> They were very different in almost every personal way—Einstein gregarious, happy, full of laughter and common sense, and Gödel extremely solemn, very serious, quite solitary, and distrustful of common sense as a means of arriving at the truth. But they shared a fundamental quality: both went directly and wholeheartedly to the questions at the very center of things.[17]

As Gödel later told Einstein's biographer Carl Seelig, "Our discussions principally related to philosophy, physics, and politics. . . . I have often pondered why Einstein took pleasure in his conversations with me, and I believe one of the causes is to be found in the fact that I frequently was of the contrary opinion and made no secret about it." It was a welcome change from the awestruck veneration of most of Einstein's colleagues. "He appreciates Gödel beyond anything," Morgenstern wrote in his diary, though not without adding, "Einstein sees everything very clearly; the pathological traits and the genius."[18]

But there was also a simple human bond between the two towering intellects that affected them both deeply. When Einstein was confined to bed for several weeks in the summer of 1947, Gödel visited every week, and Einstein was equally solicitous of Gödel, when he suffered his frequent bouts of illness. They exchanged small but thoughtful gifts—a vase for flowers from Einstein as a housewarming present, a framed etching ("Quiet Hour," by the highly regarded American lithograph artist Stow Wengenroth) from Gödel to Einstein on his seventieth birthday, a wool vest that Adele hand-knitted for Einstein as a special present one time. Learning of Marianne Gödel's fascination with her son's famous friend, Einstein wrote her a touching letter that delighted her. "I can imagine that Einstein wrote nicely to you," Gödel told his mother. "He is just friendliness itself (at least to me)."[19]

With Oskar Morgenstern

Morgenstern's friendship with Gödel however would be the more enduring, as well as the deepest of both of their lives. Morgenstern certainly did not need the aura of Gödel's friendship for either professional prestige or personal self-esteem. If anything, Morgenstern was the more renowned of the two at the time they met in Princeton, with an international reputation as an economist whose counsel was sought at the highest levels of government and business. In collaboration with John von Neumann, he was developing the groundbreaking concept of game theory, which would revolutionize the understanding of strategic interactions between competitors in situations as diverse as markets, political negotiations, and nuclear deterrence. He was also a charismatic and dashing figure: his mother was said to have been the illegitimate daughter of Germany's short-lived emperor Friedrich III, and Morgenstern had inherited the charm and air of a natural aristocrat. He was an expert horseman, a sophisticated lover of music and literature, and, approaching forty and handsome and still unmarried, a magnet for a series of pretty and accomplished girlfriends not much more than half his age.

The trust and deep affection that grew up between the two men was as profound as it was sincere, uncomplicated by calculation or self-interest. "He really is an *especially* likeable person," Gödel described him to his mother, and he spoke openly to Morgenstern about his work and life as to no one else.[20] For thirty years they usually saw each other every few days, often having long telephone conversations in the intervals between.

"Why he picks me is not clear," Morgenstern wrote early on in his diary, "but apparently likes & is always especially devoted to me." Gödel reciprocated by reading and discussing Morgenstern's work with him, and after Morgenstern's marriage Gödel and Adele often celebrated Christmas and birthdays together with Morgenstern's growing family, as well as regularly visiting each other's homes for tea or dinner. Gödel described Morgenstern to his mother as "the only one we see frequently."[21]

Morgenstern reflected that he had many good acquaintances, but no true friend other than Gödel. "*Was für ein Kopf!*" he wrote of Gödel in wondering admiration—"What a brain!"—and looked upon their conversations, especially after the death of his close colleague von Neumann in 1956, as "an island in the sea of mediocrity that the university represents. Every time a suggestion, a thought; always so friendly, the cheerful undertone and wondering about trivial things." Years later Gödel presented Morgenstern a photo of himself with the dedication, "In old friendship"—"which it really is," Morgenstern agreed. "His friendship is as touching as it is amazing."[22]

"I like him immensely," Morgenstern confided to his diary, "& nobody, none of my friends inspires me as much as he does."[23]

STRANGE DOINGS IN THE UNIVERSE

"That he is an important man is shown again and again," Morgenstern observed shortly after Gödel arrived in Princeton in 1940, "but he is a little crazy."[24]

In their first three years in Princeton Gödel and Adele moved three times, because, he told his mother, "I could not tolerate the bad air." He complained that Princeton's muggy summer weather was affecting his heart, but elaborated to his physician brother, "In my opinion, a further reason was (even though the doctors wanted to take no notice of it) the furnace in our apartment. It was supposedly a warm water heater (with circulating water), but it very often smelled of smoke in the rooms. The skilled craftsmen that I had come made fun of me of course but one of them actually admitted that perhaps some kind of fumes could be coming from the cellar through chinks in the walls or through the windows. (I also of course told the people at the Institute. I didn't have headaches.)" In one of the apartments, Gödel went so far as to have the radiators removed, which, as the Institute's director Frank Aydelotte reported, "makes the apartment a pretty uncomfortable place in the winter time."[25]

One day in the local appliance store in Princeton, Marston Morse, a mathematical colleague at the Institute, heard another story of Gödel's strange fears. Noticing where he worked, the salesman asked, "Do you know Professor 'Go-dell'?" "Yes, I know him," Marston replied. "I think he's crazy," the salesman said. "Why?" asked Marston. "I've delivered three refrigerators to him, and he thought they gave off a toxic gas."[26]

He and Adele were having constant fights with their landlords, though not all of that was due to their difficult and demanding natures: the landlords, Gödel reported, were "all on their high horses" due to Princeton's extreme housing shortage, and many refused to make even necessary repairs. To escape the heat the next two summers, the Gödels decamped for extended stays in Maine, spending two months in the summer of 1941 at the Mountain Ash Inn in Brooklin, a vacation spot by the ocean near Mount Desert Island popular with academics, where Veblen spent his summers, and returning the next year to nearby Blue Hill—where Gödel's accent and long solitary walks along the coast aroused the suspicions of locals that he was a German spy signaling to submarines. His and Adele's eccentricities also left a lasting impression on the hotel's owner, who years later remembered Adele's refusal to allow the maids in to clean the room, and Gödel's subsequent lengthy correspondence in which he accused the hotel of having stolen his trunk key.[27]

As well as the fate of his family, persistent uncertainties over his own status were preying on his mind. The Institute continued to renew his position as a temporary member at a salary of $4,000 a year, but only one year at a time, no doubt reflecting concerns over his long-term stability. "I am not exactly enthused about this year-to-year position, for in addition I am given to understand that the available funding is getting ever tighter (perhaps due to the war?)," he wrote his brother.[28] The German consulate in New York was meanwhile demanding that he formally apply to extend his leave of absence, even as Rudi reported from Vienna that the Nazi military and academic officials there kept sending letters demanding to know his whereabouts, apparently unaware that he had left for America with official approval. There was also a tangle over taxes he owed in Germany and endless complications about renting out the last apartment he and Adele had lived in during their final months in Vienna, at Hegelgasse 5.

Improbably, in the midst of it all, Gödel's appointment as Dozent of the New Order came through. The elaborate embossed diploma, dated June 28, 1940, and conferring upon him "the Führer's special protection," was still sitting uncollected in his university file along with a blank receipt for him to sign when it was rediscovered six decades later.[29]

A few months later he wrote to Rudi,

> I was very astonished that I was being inquired about by the Military Command. After all, I requested an extension of my leave of absence back at the end of April and they told me at the local consulate that it had been reported to the Vienna authorities and that the matter was completely in order. About my reappointment "in absentia" I was just as surprised. Until May 1941 I will stay here at any rate, for I did accept this position and have drawn part of the salary. By the way, it would interest me very much whether any salary (and in what amount) would be associated with my Vienna position? Could you not find that out? According to the new laws,

all dozents are to be paid. If you speak to the people at the university can you tell them too that I submitted an application for extension of leave of absence to the Vienna Rectorship and the Ministry for Science, Education, and Culture via the local consulate, primarily with the justification that 1) I have no money (no dollars) for the return trip and 2) I have no money to live on in Vienna. They apparently know nothing of this request in Vienna, otherwise they would not ask me to appear in the upcoming days.[30]

In March 1941 he again applied to extend his leave of absence, but this time the consulate made clear that the German authorities were beginning to look with suspicion upon his extended stay in America. The ministry in Berlin reported several months later:

As the German General Consul in New York recently reported, Dr. G o e d e l has sought a further leave of absence, because the Institute in Princeton has offered him a stipend of $4000 for further research work in the coming academic year. As the General Consul explained to him that his acceptance of this offer would be undesirable, but that another suitable activity could not be found, Dr. Goedel asked with a view to this to be informed, in the case a further leave of absence were denied, if following a possible repatriation he would receive a paid position at a German university. He added, however, that owing to a heart condition, strenuous administrative or teaching activities would not be possible for him.[31]

The word the official used for "repatriation," *Heimschaffung,* has the ominous connotation of being bundled off or shanghaied.

By December 1941 Gödel's deteriorating mental state so alarmed Aydelotte that he wrote to Dr. Max Gruenthal, a psychiatrist in New York City whom both Gödel and Adele had been consulting, to ask if he might be dangerous. "I am naturally a good deal concerned about Dr. Goedel's condition and should be most grateful if you could give me your general opinion of his case, letting me know especially whether you think it is all right for him to go ahead with his work, whether you consider him improving, and whether there is anything we could do here at the Institute to ease the mental tension from which he is evidently suffering. I should like also especially to know whether you consider that there is any danger of his malady taking a violent form which might involve his doing injury either to himself or to other people." Gruenthal was able to reassure him on that score at least. "I do not see any acute danger of his malady taking a violent form," he wrote back, and Aydelotte replied that he would try to have Gödel's friends encourage him to consult his psychiatrist more often than he had been doing.[32]

Adding to the anxieties of their life in the war years was that he and Adele had registered as German citizens on their arrival, which classified them as enemy aliens and meant that every time they wanted to travel to New York to attend a meeting or keep an appointment with Dr. Gruenthal, or take a vacation on the Jersey Shore, they had to write in advance to the U.S. Attorney in Trenton for permission. They were finally allowed to change their registration to "Austrians," whom the State Department were willing to recognize as victims rather than allies of Nazi Germany.[33]

As if all of those worries were not enough, in April 1943 Gödel was ordered to report to the Trenton Induction Center for an army medical exam. Though not yet a citizen (he and Adele had taken out their "first papers" in December 1940, the first step in applying for citizenship, for which they would be eligible after five years of residence), like all U.S. permanent residents he was subject to the military draft. Demonstrating if nothing else the shared incompetence of military bureaucracies everywhere, the local Selective Service Board classified Gödel as 1-A, fit for military duty, just as had its German counterpart three years earlier.

Attempting to head off a complete disaster, Aydelotte sent an urgent plea to the draft board. "Dr. Gödel, like most refugees from Nazi Germany, is eager to do anything he can in support of the American war effort," he wrote, "but under the circumstances I think I ought to inform the Selective Service Board that Dr. Gödel has twice since he has been in Princeton shown such signs of mental and nervous instability as to cause the doctors who were consulted to diagnose him as a psychopathic case. . . . . He responded so well to the treatment that we invited him to come again to the Institute in 1940 and he has been here since that time. Last year, however, the symptoms returned and it has been necessary for him again to have medical treatment." While Gödel was a mathematical genius, Aydelotte explained, "this ability, however, is unfortunately accompanied by certain mental symptoms which, while they do not prevent active work in mathematics, might prove serious from the standpoint of the Army." Aydelotte offered to have Dr. Joseph S. Vanneman, the Princeton University physician and chief physician at Princeton Hospital, who had treated Gödel on both occasions, provide additional details about the case.[34]

The draft board replied that while sympathetic to Aydelotte's "knowledge of Mr. Gödel's condition," they could do nothing except forward the information to the induction center with a request that the army's own doctors conduct a "complete psychiatric examination." In the end Gödel did report for his required induction exam—his *Assentierung* he called it, using a quaint Austrian term from the days of the Austro-Hungarian Empire—and was accompanied to Trenton by a fellow professor who had received a draft notice at the same time. For whatever reason, he was mercifully reclassified 4-F, permanently excused from military service.[35]

Gödel's ability to carry on doing mathematical work despite his periodic psychological crises was, as Aydelotte pointed out, a striking aspect of his personality, and the latest episode was no exception. Not long after his recovery he was plunged full tilt into an entirely new area of investigation, which led him to a highly original solution to the equations of Einstein's Theory of General Relativity. It was a mathematical tour de force.[36] But what interested Gödel even more were its strange philosophical implications, particularly concerning the nature of time.

As Gödel explained in a short essay he was asked to contribute to a volume in the *Library of Living Philosophers* series devoted to Einstein, "One of the most interesting aspects of relativity theory for the philosophical-minded consists in the fact that it gave new and surprising insights into the nature of time, of that mysterious and seemingly self-contradictory being which, on the other hand, seems to form the basis of the world's and our own existence."[37] It was well known that even in Einstein's more elementary theory of special relativity, observers moving at different speeds not only experience the passage of time differently, but even have different perceptions of whether two events occur simultaneously. But there remains in the standard cosmological solutions to Einstein's equations a "privileged," and thus in a sense objective, "cosmic time" that corresponds to what an observer moving along with the average mass of the expanding universe would perceive.

Gödel's solution, however, eliminated even that toehold on the objective lapse of time. In Gödel's model, the universe is not expanding, but instead matter everywhere is rotating, along the same parallel axis, about its local inertial frame of reference. In his universe there is no place that is distinguishable from any other as a universal time standard. Gödel pointed out some "other surprising features" of the world he had derived from Einstein's equations:

> Namely, by making a round trip on a rocket ship in a sufficiently wide curve, it is possible in these worlds to travel into any region of the past, present, and future, and back again, exactly as it is possible in other worlds to travel to distant parts of space.

As Gödel readily admitted, "This state of affairs seems to imply an absurdity. For it enables one, e.g., to travel into the near past of those places where he has himself lived. There he would find a person who would be himself at some earlier period of his life. Now he could do something to this person which, by his memory, he knows has not happened to him." But, he countered, this possibility for cosmic mischief did not automatically rule out his model. A calculation of the energy required to make such a voyage suggested it might be barred as a practical matter; the spaceship would have to travel at a minimum of two-thirds the speed of light, and to complete the journey in the lifetime of its occupants would need to carry fuel that weighed something over a billion billion times as much as the weight of the ship itself. Thus the reason such time-travel paradoxes do not occur might be not because his model was wrong, but only because they are forever technologically out of reach.[38]

Gödel's model faced the simpler objection that it does not comport with the strong observational evidence, provided by the red-shift of distant galaxies, that the universe is indeed expanding. But Gödel argued that the deeper implication of his finding was that if such a world can exist at all, it renders meaningless the notion of the passage of time in any objective sense. Time thus becomes not a physical reality, but merely the product of each observer's perception, an argument akin to what Immanuel Kant and other "idealist" philosophers had maintained. "If the experience of the lapse of time can exist without an objective lapse of time," Gödel concluded, "no reason can be given why an objective lapse of time should be assumed at all."[39]

His friend Georg Kreisel thought he perceived "between the lines" of Gödel's essay a hint of his fascination with the idea of ghosts that could reappear from the past, and of a timeless future where the souls of men might live on.[40]

CITIZEN GÖDEL

In the final months of the war Vienna had come within range of American bombers operating from Italy and Soviet forces closing in from the east. The city's oil refineries were repeatedly targeted in raids that sent errant bombs raining onto the city center, causing heavy damage to the Opera House and Burgtheater, filling the Ringstraße with avalanches of debris, and in one attack that missed its target altogether striking the famous zoo in the grounds of Schönbrunn palace, killing thousands of animals. The university's main building was hit by twenty-six bombs during the raids.

In the summer of 1945 Gödel received the first word from his family in almost a year. He had received some letters from them via the Red Cross during the war, but all of his sent to them had gone astray. "Dear Mama and dear Rudi," he wrote a month after the final end of the war, "I am happy to hear again from you in more detail after such a long time and that you have managed to survive the past months. . . . We will of course send packages right away as soon as it is possible. . . . Might it be possible as early as next summer for us to come visit in

Vienna? Hopefully conditions will improve quickly and we will hear from you very soon."[41]

They sent a blizzard of food parcels and CARE Packages to Vienna for them and their other relatives, along with vouchers from Vienna's famous coffee supplier, Julius Meinl, that could be redeemed in Vienna for staples like flour, dried milk, and instant coffee and small luxuries like apricot jam, sardines, croissants, ham, and cigarettes, shipped directly from America.

Slowly the news began to filter back about what had happened during the war and the fate of their friends and relations. It took three letters from his mother telling him of Fritz Redlich's death in a concentration camp for Gödel to accept the reality—in retrospect, a disturbing harbinger of his growing tendency to resist tragedy with irrelevantly rigid logic. "I remember now that you had already written about Redlich," he wrote back to her in 1947, "but these things didn't stick in my memory because I couldn't quite believe it. Redlich was a half-Jew after all and I scarcely believe that there were gas chambers for them (they were even subject to military service)."[42]

Brno had suffered the heaviest air attacks of any Czech city, with 30,000 apartments destroyed in Allied bombing raids aimed at the city's armaments factories. All the torments of Nazi brutality and retaliation added to the horrors of the war years and their immediate aftermath. The Germans in Brno, like the rest of the Sudeten Germans, had been the most ardent Nazis in the entire Reich, and Rudi had been deeply worried that his mother's often indiscreet remarks about Hitler and National Socialism would land her in serious trouble. Frequent Nazi executions of Czech "traitors," announced by red posters on the streets, added to the now implacable hatred of the Czechs for their German overlords, and relations between Gödel's mother and her Czech housekeeper became extremely strained. On a visit to Vienna in August 1944, she had been prevailed upon by Rudi not to go back again, and moved into the small bachelor apartment adjoining his surgery on Lerchenfelder Straße. The final three days of the war they had spent in the building's cellar while the Russians battled with the last German holdouts. They returned from their first walk around the devastated city, its streets filled with mountains of shells and unburied corpses, in tears.[43]

But Marianne's move to Vienna spared her from the furious reprisal of the liberated Czechs unleashed by the war's end, and almost certainly saved her life. On May 12, 1945, a week after the final German surrender, the returning Czech President Edward Beneš came to Brno and declared, "The German question in the Republic must be liquidated." Twenty thousand Germans in Brno were rounded up and forcibly marched by a hastily assembled guard of Czech factory workers, partisans, and soldiers to the Austrian border. The Austrian provisional government created under the Soviet military occupation barred their entry, leaving thousands to be thrown into makeshift camps, while the rest made their way back on foot to Brno. Some 1,700 died from starvation or outright murder along the way.[44]

The new Czechoslovak government confiscated German property across the country, seizing the Gödels' villa along with the rest. Although Gödel would try for years to recover the property, engaging a local lawyer and unsuccessfully seeking the aid of the U.S. State Department, he acknowledged that the Czech retaliations against its former German citizens "are probably only explicable by how Hitler dealt with the occupied territories," and he was later repelled by the enduring German-nationalist overtones in books and newspapers published by exiled Brünnites living in Germany in the 1950s.[45]

But Marianne's greatest problem now lay in Vienna. In January 1946 Rudi wrote with the alarming news that the Russians had begun to round up Germans from Czechoslovakia who, like their mother, had never actually received Austrian citizenship:

> The expulsion from here has nothing to do with the [Nazi] Party, but rather affects all *Volksdeutsche* from the Czechoslovak Republic, thus even Mama who was not a Party member! I will here naturally also put all levers in motion to keep Mama here, especially with the point that she lived for many years at the time in Vienna. . . . If it were possible for you to get from the authorities at Princeton University a letter in which the local military gov-

ernment is requested out of regard for your position to make an exception here for Frau M. Gödel, that naturally would be an even better possibility.[46]

In the end their fears proved unfounded. But Adele was equally worried about her family—"completely unhinged" about the situation in Vienna, Morgenstern reported—and was determined to visit at the first opportunity. It took repeated trips and phone calls to the Austrian mission in Washington and its New York consulate ("one notices immediately that one is once again dealing with those dear agencies of the homeland," Gödel drolly noted) to secure all the necessary permissions. Adele departed for Vienna on May 2, 1947, and returned on December 4, just over seven months later—a "marriage vacation," she afterward termed their time apart.[47] During her absence Gödel's friends took turns solicitously looking after him, feeding him meals, taking him on excursions to the country, helping him find a housemaid who could cook and clean for him. "Morgenstern is very nice and makes an effort to provide me with diversions," he reported to his mother.[48]

Adele's visit did little to improve his family's attitude toward her, which was destined to begin with a frost and end with a frost. Interviewed in the 1990s, more than a decade after both Kurt's and Adele's deaths, all Rudi would offer when asked about Adele was the stiff comment, "I would not presume to pass judgment on my brother's marriage."[49] All of Marianne Gödel's letters to her son in America were destroyed after his death, but his replies make clear the tensions between the two women, as well as his own enduring ambivalence about life with Adele.

These family dynamics were inescapably intertwined with Gödel's own psychopathologies, which in their most salient aspects could have been a textbook study for classical psychoanalytic theory: the coldness he felt toward his father and the intimate emotional ties he maintained throughout his adult life with his mother; his anal obsessiveness; his perpetually juvenile tastes in art, music, and decorations, and fixation on childhood toys, games, and fairy tales; his sexual attraction to domineering older women. His letters to his mother, which he dutifully wrote every other week, then once a month, from the end of the war on, usually on Sunday evenings, show a genuine kindness, family closeness, and thoughtful consideration; he told her charming details of his and Adele's house and garden in Princeton and their vacations to the shore, patiently explained his work and satisfied her curiosity about Einstein, shared his thoughts and insights about books and plays, politics and politicians, and the meaning of life, all with an unaffected ingenuousness and a striking absence of self-importance: he never played the great man. He also regularly gave her reports on the state of his bowels, and exactly what Dr. Freud would have made of a fifty-four-year-old man being sent an enema tube by his mother to try out, as she did on one occasion, is another matter altogether.[50]

He wrote to his mother a few weeks after Adele's arrival in Vienna, dismissing her apparent insinuation that Adele was neglecting her wifely duties by leaving him to fend for himself for so long. "I usually eat in a restaurant and the little bit of dishes that I need I can wash up with hot water myself right after the meal and let it dry on a rack; it does not take much time. Making beds is a healthy gymnastic exercise and I do not have anything else to do after all," he insisted. At the same time he frankly acknowledged Adele's serious problems:

> Regarding Adele's "bad mood" toward you, I must unfortunately say that Adele seems to be not quite normal in her relationship to other people. She is often deadly insulted about things that a normal person would not even take notice of or which are certainly not ill-intended, possibly even well-intended. Also her dislike toward the landlord of our apartment completely exceeds the boundaries of normalcy, even if she possibly does justifiably complain about some things. In general she often has, in my opinion, completely false ideas about other people, especially in the sense of some animosity directed toward her. Aside from that, here she has clearly shown symptoms of hysteria (in the sense of an *illness*, not just stubbornness) and probably belongs in the care of a doctor for nervous disorders if only it were not so expensive and success

so uncertain. I certainly hope that being with her mother and sister will have a positive effect on her and merely wanted to inform you just in case. Please of course do not tell her under any circumstances anything of what I am writing you here; that would only make matters worse.[51]

Morgenstern made his own trip to Europe that summer and took the opportunity to call on Gödel's family, charming Marianne and getting on well with Rudi ("a very nice & charming man 45; much like his brother. Excellent taste, fine library. The mother is nice too") and astonishing them both by explaining to them, clearly for the very first time, the renown and importance of their self-effacing son and brother. "They all did not really understand who Kurt really is. K never told them, in Vienna he is largely unknown & it pleased them greatly to hear from me."[52]

Shortly after Adele's return they both went in turn to Trenton for their citizenship examinations, the last hurdle in becoming American citizens. Gödel's sponsors were Einstein and Morgenstern, and he was to appear for his examination before the same the federal district judge who had presided over Einstein's naturalization ceremony seven years earlier, Phillip Forman.

The story of Gödel's examination has entered the realm of legend and been told in many embroidered versions, but Morgenstern left a firsthand account that is surely the most accurate. Months before his citizenship test Gödel had commenced upon an exhaustive study of American history, government, current events, laws, and statistics, filling pages of notes in Gabelsberger script about American Indian tribes, the names of British generals in the Revolutionary War, the National Bankruptcy Act of 1863, the office of the Postmaster General, and checking out from the Princeton University Library the New Jersey Revised Statutes, the 1901 Sanitary Code, and the Act of Incorporation of the Town of Princeton. Every few days he would call Morgenstern to ask for more books, and to pepper him with questions about what he had so far discovered. "Gödel reads the World's Almanac & calls me many times, amazed at the facts he finds & those that he expects to find & are not in it he attributes to evil intent to hide them. Harmless-funny. It amuses me very much," Morgenstern wrote.[53]

As the date approached he began obsessively interrogating Morgenstern about the organization of local government. Morgenstern recalled,

He wanted to know from me in particular where the borderline was between the borough and the township. I tried to explain that all this was totally unnecessary, of course, but with no avail. . . . Then he wanted to know how the Borough Council was elected, the Township Council, and who the Mayor was, and how the Township Council functioned. He thought he might be asked about such matters. If he were to show that he did not know the town in which he lived, it would make a bad impression.

I tried to convince him that such questions were never asked, that most questions were truly formal and that he would easily answer them; that at most they might ask what sort of government we have in this country or what the highest court is called, and questions of this kind.

At any rate, he continued with the study of the Constitution. He rather excitedly told me that in looking at the Constitution, to his distress he had found some inner contradictions and that he could show how in a perfectly legal manner it would be possible for somebody to become a dictator and set up a Fascist regime, never intended by those who drew up the Constitution. I told him that it was most unlikely that such events would ever occur, even assuming he was right, which of course I doubted. But he was persistent and so we had many talks about this particular point. I tried to persuade that he should avoid bringing up such matters at the examination before the court in Trenton, and I also told Einstein about it: he was horrified that such an idea had occurred to Gödel, and also told him he should not worry about these things nor discuss that matter.

On the day of the examination, Morgenstern picked Gödel up in his car, then drove to Einstein's home and the three of them set off to Trenton. Einstein, with his usual love of mischief, turned around to Gödel in the back seat and asked sternly, "Now, Gödel, are you *really* well prepared for this examination?"—which to Einstein's glee had exactly the effect intended, sending Gödel into a momentary panic.

Entering the courtroom, Judge Forman was delighted to see Einstein and chatted with his famous visitor briefly before turning to Gödel.

"Now, Mr. Gödel, where do you come from?"

"Where I come from? Austria."

"What kind of government did you have in Austria?"

"It was a republic, but the constitution was such that it finally was changed into a dictatorship," Gödel replied.

"That is very bad," the judge said. "Of course that could not happen in this country."

"Oh, yes," Gödel exclaimed. "I can *prove* it!"[54]

Forman, Einstein, and Morgenstern immediately joined in shutting Gödel up before he could say anything further about his pet idea, and the rest of the ceremony went off without incident.

(No one has found in Gödel's papers an explanation of what the flaw was he claimed to have discovered, but one imaginative legal scholar has plausibly suggested that Gödel might have observed that while the Constitution makes it deliberately difficult to enact amendments, there is nothing to prevent the amendment process itself from being amended to weaken those safeguards—a pleasingly "Gödelean" bit of self-referential logic.)[55]

The Institute meanwhile had finally made his appointment permanent; the new position came with a 50 percent raise, to $6,000 a year. Again, it was von Neumann who pulled strings to bring it about, indignantly remonstrating to Veblen, "Gödel did some of his best work (continuum hypothesis) at the Institute—actually at a time when he was less normal than now. The Institute is clearly committed to support him, and it is ungracious and undignified to continue a man of Gödel's merit in the present arrangement forever."[56]

His citizenship and permanent position cemented the feelings he had already conveyed to his mother of his happiness and security in his new home. "I do feel very at home here in this country," Gödel told her, "and would not go back to Vienna even in the event that I were to be offered something. Aside from all personal circumstances, I find the country and the people here 10 times more appealing than at home."[57]

Plato's Shadow

METAPHYSICAL AXIOMS

At the beginning of 1951 the American Mathematical Society selected Gödel for a signal honor, inviting him to deliver the Josiah Willard Gibbs Lecture at its annual meeting, being held that year in Providence, Rhode Island, in December. Intended to showcase for the public the greatest achievements of mathematics, the Gibbs Lectures had previously featured such giants of the field as Einstein, von Neumann, Hermann Weyl, the English number theorist G. H. Hardy, and the theoretical aerodynamicist Theodore von Kármán. Gödel's lecture would be the first given by a logician.

He almost didn't make it. On February 1, he summoned to his house the local doctor, Joseph M. Rampona, whom he had been consulting since his first years in Princeton. When Rampona arrived he was vomiting blood. The diagnosis of a bleeding ulcer was unmistakable. But, as Rampona observed, recalling not only this but many other similar encounters, "He was a patient with whom it was rather difficult to deal."

"What do you think I have?" demanded Gödel.

"You have a stomach ulcer," Rampona replied.

"I don't believe it," Gödel said.

Rampona finally had to get Einstein to come over and convince his difficult patient that he indeed had to go to the hospital, without delay.[1]

"He thinks he's going to die & it's hard to talk him out of things like that," Morgenstern wrote after visiting him in the hospital, where he had insisted on dictating his will to his friend. He was receiving blood transfusions, which almost certainly saved his life, but all the while kept insisting that the doctors at Princeton Hospital did not know what they were doing and refused to follow their orders or take the medications they prescribed. After two weeks, he scribbled a frantic note in German to the Institute's new director, J. Robert Oppenheimer, saying he had to speak to him urgently about being moved to a different hospital. "I am getting steadily worse and worse here and 2 gross errors in treatment have already occurred," he wrote. "If I am ever to regain my health, it certainly will not be here."[2]

Oppenheimer, perhaps wondering exactly how he had gone from directing the atomic bomb program to playing nursemaid to a semi-hysterical logician, wrote back what he hoped was a soothing reply: "I talked to your doctor about the worries expressed in your letter to me. He tells me that you will be leaving the hospital very soon. I have confidence that he understands your troubles and will take good care of you. Get well soon."[3]

As with all of his previous maladies, self-generated, imaginary, and otherwise, Gödel recovered his health and equanimity almost as quickly as he had plunged precipitously close to death. By the beginning of March he wrote Rudi a matter-of-fact account of his illness in which he even made a small joke about it: "I wrote you once that everything here is so precisely regulated that the winter cold or the summer heat begins exactly on the 1st of the month. With illnesses it therefore also seems to be the case."[4]

While he was in the hospital, Oppenheimer asked Morgenstern if there was anything that could be done that would make Gödel particularly happy. "I said he should finally be appointed professor," Morgenstern said. But Institute politics, not for the first time, at once threw up an obstacle, with several members of the mathematics faculty strongly objecting on both personal and procedural grounds.[5]

Two weeks later Oppenheimer had another idea: the first Albert Einstein Award was to be presented on March 14, Einstein's birthday; it carried $15,000 and a gold medal, funded by Admiral Lewis Strauss, a wealthy businessman and member of the Institute's board who had gone on to become a key force in the development of nuclear weapons and nuclear energy as a member of the Atomic Energy Commission. The award committee had already selected Julian Schwinger, a young theoretical physicist, to receive the prize. Oppenheimer was worried it might be embarrassing to reverse course at this point, even though a formal announcement of the winner had not yet been made, and was concerned, too, about the appearance of cronyism in giving the prize to Gödel because his close friends von Neumann and Einstein were both on the selection committee. But a plan to divide the award between the two men was quickly agreed to.[6]

Despite some last-minute trouble from Adele, who a few days before

Einstein Award ceremony, 1951, with Admiral Strauss and Julian Schwinger

the ceremony started insisting that she had to be seated next to her husband, all went off without a hitch ("she behaved very well," reported Morgenstern). Von Neumann read a glowing tribute, and despite Gödel's modest insistence to his mother, "I hate those sort of things"—joking that von Neumann "really praised me to the heavens, so that every unbiased listener would have had the complete impression that the prize was not adequate to my accomplishments"—he was visibly elated when Einstein personally presented him the award, at a luncheon for seventy-five guests hosted by Strauss at the Princeton Inn. To the wonder of all assembled, Morgenstern noted, "Einstein even had a tie on! He was in the best of spirits and when he gave Gödel the medal he said: 'And here my dear friend for you. And you don't need it!' "[7]

The award was noted on the front page of the *New York Times* and in major Austrian newspapers, and Gödel received congratulations from his old school friends Fritz Löw Beer and Harry Klepetar. In June he was awarded an honorary degree by Yale University on the occasion of the 250th anniversary of its founding. He again reported the news to his mother with a tone of boyish ingenuousness: "There were ca. 25 honorary doctorates awarded to deserving scientists. I was by far the youngest; all the others were 60 or even older. For the ceremony I had to wear an academic robe and a beret which Adele found very flattering."[8]

He was fully recovered by the time he and Adele traveled to Providence to deliver his Gibbs Lecture the day after Christmas. In what was fated to be his last public presentation, he offered the clearest and most forceful explanation of his views on Platonism, the nature of mathematics, and mathematical truth. Gödel would later be irked by Bertrand Russell's jokey dismissal of him as an "unadulterated Platonist," but Russell had a point. Describing the weekly conversations he had had with Gödel and Wolfgang Pauli at Einstein's house during an extended visit to Princeton in 1944, Russell wrote in his autobiography:

These discussions were in some ways disappointing, for, although all three of them were Jews and exiles and, in intention, cosmopolitans, I found that they all had a German bias towards metaphysics, and in spite of our utmost endeavours we never arrived at common premises from which to argue. Gödel turned out to be an unadulterated Platonist, and apparently believed that an eternal "not" was laid up in heaven, where virtuous logicians might hope to meet it hereafter.[9]

In an unsent reply to a query from the archivist of Bertrand Russell's papers at McMaster University many years later, Gödel irritatedly cited Russell's own words in his *Introduction to Mathematical Philosophy* against him. "I have to say *first* (for the sake of truth) that I am not a Jew (even though I don't think this question is of any importance)," Gödel wrote. "Concerning my 'unadulterated' Plat[onism], it is no more 'unal.' than Russell's own in 1921," an allusion to a passage in Russell's *Introduction* where he wrote, "Logic is concerned with the real world just as truly as zoology." Gödel had pointedly quoted that same passage in an essay he contributed to a 1944 volume devoted to Russell in the *Living Philosophers* series.[10]

In the Gibbs Lecture, Gödel marshalled several powerful arguments against the notion that mathematics is nothing but a creature of man's own invention. To begin with, the very act of mathematical creation "shows very little of the freedom a creator should enjoy," as the creature almost immediately begins to take on a life of his own and goes on to defy and make demands on his creator in other ways:

> Even if, for example, the axioms about integers were a free invention, still it must be admitted that the mathematician, after he has imagined the first few properties of his objects, is at an end with his creative ability, and he is not in a position also to create the validity of the theorems at his will. . . .
>
> If mathematical objects are our creations, then evidently integers and sets of integers will have to be two different creations, the first of which does not necessitate the second. However, in order to prove certain propositions about integers, the concept of [a] set of integers is necessary. So here, in order to find out what properties

we have given to certain objects of our imagination, [we] must first create certain other objects—a very strange situation indeed![11]

He also drew a striking connection between his incompleteness results and these philosophical implications about the nature of mathematics. Both of his Incompleteness Theorems proved that no finite process of inference from axioms within a well-defined system can capture all of mathematics. But that, Gödel pointed out, leads to an interesting either–or choice: either the human mind *can* perceive evident axioms of mathematics that can never be reduced to a finite rule—which means the human mind "infinitely surpasses the powers of any finite machine"—or there exist problems that are not merely undecidable within a specific formal system, but that are "absolutely" undecidable.[12]

Both choices point to a conclusion "decidedly opposed to materialistic philosophy," Gödel observed. If the mind is not a machine, then the human spirit cannot be reduced to the mechanistic operation of the brain, with its finite collection of working parts consisting of neurons and their interconnections. If, however, the mind *is* nothing but a calculating machine, then it is subject to the limitations of the Incompleteness Theorem, which leads to the thorny fact that numbers possess at least some properties that are beyond the power of the human mind to establish: "So this alternative seems to imply that mathematical objects and facts (or at least *something* in them) exist objectively and independently of our mental acts and decisions, that is to say some form or other of Platonism or 'realism' as to the mathematical objects."[13]

He acknowledged that owing to "the undeveloped state of philosophy in our days," such arguments cannot be asserted with mathematical rigor: they are not "a real proof" for mathematical Platonism. But Gödel suggested that developments in foundations of mathematics, his Incompleteness Theorems in particular, at least lent strong support for this conclusion, and disproved the view that "considers mathematics to consist solely in syntactical conventions and their consequences."[14]

He ended by quoting the nineteenth-century mathematician Charles Hermite:

There is, if I am not mistaken, a whole world which is the set of mathematical truths, into which we gain access only through intelligence, just as there is a world of physical realities; the one and the other independent of us, both of a creation divine.[15]

Gödel remarked to Morgenstern a few years later, "Philosophy today is—at best!—where the Babylonians were with mathematics." But he was firmly convinced that one could axiomatize philosophy just like mathematics, and thus apply the objective mechanism of logic to explain fundamental questions of human existence. "One could establish an exact system of postulates employing concepts that are usually considered metaphysical: 'God', 'soul', 'idea'," he had explained to Rudolf Carnap when they had met for a lengthy philosophical discussion in Princeton in 1940. "If this is done accurately, there would be no objection." And as he told Hao Wang, "The beginning of physics was Newton's work of 1687, which needs only very simple primitives: force, mass, law. I look for a similar theory for philosophy or metaphysics. Metaphysicians believe it possible to find out what the objective reality is; there are only a few primitive entities causing the existence of other entities."[16]

In one of the many notebooks in which he recorded his philosophical aperçus, he wrote a list of his basic precepts which underscored how little his beliefs fit into the twentieth-century world of scientific empiricism, or for that matter the Age of Enlightenment altogether. Like his philosophical hero Leibniz, he explained, "My theory is rationalistic, idealistic, optimistic, and theological," committed to accessing the immaterial world of higher philosophical truths through the power of sheer abstract logical reasoning.[17]

1. The world is rational.
2. Human reason can, in principle, be developed more highly (through certain techniques)
3. There are systematic methods for the solution of all problems (also art, etc.)

4. There are other worlds and rational beings, who are of the other and higher kind.

5. The world in which we now live is not the only one in which we live or have lived.

6. Incomparably more is knowable a priori than is currently known.

7. The development of human thought since the Renaissance is thoroughly one-sided.

8. Reason in mankind will be developed on every side.

9. The formally correct is a science of reality.

10. Materialism is false.

11. The higher beings are connected to other beings by analogy, not by composition.

12. Concepts have an objective existence (likewise mathematical theorems).

13. There is a scientific (exact) philosophy (and theology) (this is also most fruitful for science), which deals with the concepts of the highest abstractness.

14. Religions are for the most part bad, but not religion.[18]

Nothing, Gödel believed, happened without a reason. It was at once an affirmation of ultrarationalism, and a recipe for utter paranoia.

HIDDEN MEANINGS

Gödel's increasing absorption in Leibniz and philosophy dismayed more than a few of his mathematical colleagues, who did not hide from him their disappointment that he seemed to be squandering his genius on trivialities. "I always argued with him," said the mathematician Paul Erdös. "I told him, 'You became a mathematician so that people should study you, not that you should study Leibniz.'"[19]

Worse, he had several obsessive conspiracy theories about Leibniz and the fate of his philosophical writings that appalled Menger, Morgenstern, and others who heard him discourse on them. Gödel's logical insistence that nothing happens without a reason was often compounded by a very Austrian corollary: the real reason is never the apparent one. "I am particularly interested in forgeries," he told his mother, and spun fantastic theories to her and to his friends purporting to find shadowy political motives, hidden meanings, or mystical significance in things large and small: the deaths of Senator Robert Taft, Josef Stalin, and the chief justice of the U.S. Supreme Court within six months of one another in 1953 ("the likelihood of that is 1:2000"); the number of years between the First and Second World Wars; the incorrect listings of movies shown on television ("one has the impression it is sabotage"); the "suspicious" explanation given for a trolley accident in Vienna ("but the public of course must not learn that"); FDR's death just before the first meeting of the United Nations in San Francisco ("as if a secret power had taken issue with his further plans and declared: This far and no further").[20] He compulsively read books about the suicides of Prince Rudolf and King Ludwig II of Bavaria, always scorning the "official" view of their deaths, and told his mother that her seeing at the villa of the composer Franz Lehár in Bad Ischl an old theater flyer from Brünn, which listed the very same cast of a performance she had seen years earlier, was no coincidence: "You may not know that serious science has begun lately to occupy itself with such things and has invented the lovely word 'synchronicity' for it."[21]

Besides sharing his belief that nothing happens without a reason, Leibniz was made to order for a part in Gödel's conspiratorial fantasies: Leibniz kept his most important work on logic and metaphysics to himself, publishing only the shallow, reassuring, and orthodox philosophy—"this is the best of all possible worlds"—that Voltaire parodied in his character Doctor Pangloss in *Candide*.[22] But Gödel was insistent that a conspiracy was *still* afoot to suppress Leibniz's writings, and claimed to have found certain passages in his work that revealed he had discovered the paradoxes of set theory, Helmholtz's theory of resonance, the law of conservation of energy, the principle of the center of gravity, even the foundations of Morgenstern's and von Neumann's game theory, centuries ahead of others.[23]

As usual, there was no arguing him out of his ideas. Menger once tried. "You have a vicarious persecution complex on Leibniz' behalf," Menger teasingly told him, but then asked who would even have an interest in destroying Leibniz's work.

"Naturally those people who do not want man to become more intelligent," Gödel said.

Wouldn't it have been more likely, Menger countered, for the authorities to destroy the work of a radical freethinker like Voltaire?

With magnificently self-sealing logic, Gödel replied, "Who ever became more intelligent by reading the writings of Voltaire?"[24]

Gödel even entertained the idea of visiting Hannover, Germany, where Leibniz's archive was, to examine his papers firsthand. Morgenstern tried to help arrange travel funding, but also thought it "a pity that he involves himself in such fantasies," all the more so when one day he came across Gödel poring over a pile of his own unpublished work in mathematics and logic.[25]

Important & strange development: Yesterday saw Gödel at the Institute in his bare room with many notebooks (in Gabelsberger shorthand!). What is this? Oh, work, before he started on Leibniz. Math or logic? Results in either? Oh, yes; e.g. a second proof of undecidability; this time over a (new) polynomial. This time one cannot even see how to force a decision by expanding the axioms (e.g. of set theory). Just a plausibility, but that is very weak. Why doesn't he publish it? "The sales of my brochure on the Continuum Problem don't show that there is much interest in such questions!" (!!) Also the few who are interested in this could work it out for themselves! He also had still completely different results: possibly he will publish this sometime; and I should not mention anything about it. . . .

Today spoke again with Gödel. . . . The thing goes on: He has a method for finding undecidable propositions!! Only he considers it "negative"; but he will possibly nonetheless publish it all.[26]

In fact he never did publish any of it. "It is a pity that everything he learns and thinks does not go beyond himself," Morgenstern wrote a few years later. "But maybe he'll surprise us again." Gödel's friends saw all too clearly that the unlimited freedom the Institute gave its members to pursue their interests was both a blessing and a curse in his case, but were helpless to do anything about it. "Gödel is too alone; he should be given teaching duties; at least an hour a week," Morgenstern thought.[27] Menger later opined that the Institute and Princeton were in some ways exactly the wrong place for him:

> At no time in his life did Gödel need intellectual stimulation to conceive and develop original and unexpected ideas. But he needed a congenial group suggesting that he report his discoveries, reminding and, if necessary, gently pressing him to write them down. All this he *had* at the beginning of his stay in Princeton. . . . and he presumably could have found such support later. But apparently he never looked for it, and no one seemed to volunteer. The fact is that I could not observe anything of the sort in the 1950s. Rather, it soon became clear to me that he wrote up many brilliant ideas only for his desk drawer if at all. From the point of view of the outside world, his incomparable talent was lying lamentably fallow.[28]

Gödel described politics as his sole "hobby," and devoted considerable time and energy in the 1950s to reading about current affairs and applying his talent for conspiratorial interpretations there, too. In Austria he had been a Social Democrat and his own political opinions were "thoroughly anti-nationalistic."[29] A great admirer of FDR, he distrusted Truman's hardline policies toward the Soviet Union and in 1948 apparently cast his first vote as a new citizen for the third-party candidacy of the Progressive Party's Henry Wallace. "If you say it is good that the Americans possess the power," he wrote his mother, "I would sign off on that unconditionally only in the case of Roosevelt's America."[30]

The Korean War affirmed his enmity for Truman, and his fears of what lay ahead. "In your last letter you praise America to the skies. That was *my* hobbyhorse up to now, but it seems that we are destined to always be of opposite opinions. Here, gradually everything is being shifted to an economy of war," he wrote a month after the outbreak of the fighting in Korea in June 1950.[31] Two of his letters to his mother in the ensuing months sharply criticized American foreign policy, thereby arousing the scrutiny of the U.S. military censors in Vienna, who were still routinely opening and reading mail. "The political situation has continued to develop wonderfully here and one hears nothing but defending the nation, conscription, tax increase, inflation etc.," Gödel wrote in one of his suspiciously "pro-Russian" letters. "I believe myself that even in the blackest (or brownest) Hitler-Germany it wasn't that bad. . . . I hope the Germans will not be *so* stupid as to let themselves be used as cannon fodder against the Russians." A report went to FBI Director J. Edgar Hoover quoting the offending passages. But Gödel was apparently not deemed a sufficient security threat to warrant further investigation.[32]

While regarding Senator Joseph McCarthy, the anti-Communist Republican demagogue, as "roughly the American Hitler," and the Republican Party as "reactionaries," Gödel's disenchantment with what he called Truman's "warmongering" led him to turn against the Democratic Party and their nominee Adlai Stevenson in the 1952 election.[33] That year he received another honorary degree, this time from Harvard—as "Discoverer of the most significant mathematical truth of this century"—but complained afterward that he had to share the stage with Truman's secretary of defense, Robert A. Lovatt: "Through no fault of my own, I therefore, managed to get into a highly warlike company."[34]

Ernst Straus ran into Einstein one day a few months after the 1952 elections.

"You know, Gödel has really gone completely crazy," Einstein told him.

"Well, what worse could he have done?" asked Straus.

"He voted for Eisenhower."[35]

Gödel continued to support Republicans for president, even voting reluctantly for Richard Nixon in 1960 over John F. Kennedy. But he had joined the other members of the Institute in a public statement of support for Oppenheimer when Admiral Strauss, attacking him as a Communist, led the charge to have him stripped of his security clearance by the AEC and then tried to have him fired as the Institute's director. Later, upset by the Vietnam War, Gödel switched back to voting for the Democratic Party. But he remained fascinated by Eisenhower to the point of hero worship, crediting him with ending the Korean War and defusing tensions with the Communist regimes, enthusiastically telling his mother, "A man like Eisenhower at the top really only happens once every couple of hundred years." Once while talking politics with Paul Erdös he mistakenly said "Einstein" for "Eisenhower"—which they both laughingly agreed was a mix-up that the former "would certainly not have approved of."[36]

SNOW WHITE, LADY WRESTLERS, AND A FAMOUS FLAMINGO

In 1953 Gödel was finally made a professor of the Institute, thanks again in part to heavy lobbying by his protectors, von Neumann and Morgenstern in particular. "How can any of us be called professor when Gödel is not," von Neumann insisted to his friend the mathematician Stan Ulam. Much of the internal opposition had apparently come from Carl Siegel, who wryly objected, "One crazy man on the faculty is enough!"—meaning himself.[37] But Siegel had left the Institute in 1951, as had Hermann Weyl, who had also raised doubts about Gödel's mental state. The more substantive objection of others was that, as a member of the faculty, Gödel would have to take on his share of administrative duties, in particular reading applications and voting on each year's temporary members, and there were concerns whether Gödel would be willing to assume the added responsibilities.

In the event, the trouble proved the exact opposite. Gödel was so conscientious in his new duties that he delayed decisions for months as he meticulously scrutinized each candidate's publications and application. "One can make it easy on oneself by simply saying yes to everything which anyway is almost always the result of the deliberations," he told his mother. "But I do want to get oriented regarding all these things that are being discussed." His Institute colleague Atle Selberg recalled that it was often very difficult even to get applicants' files "out of his hands," much less to get an opinion out of him.[38]

The often venomous Institute politics he was inevitably drawn into further added to the burden of decision-making for Gödel. The mathematicians had a reputation as the chief troublemakers at the Institute, perpetually feuding with the director in a contest of policy and power. The conventional explanation, as one Institute member put it, was that mathematicians' work is so intense that they can only devote a few hours to it each day, which leaves them "the rest of the day to bug other people." A huge battle broke out several years after Gödel's appointment, involving the mathematicians' vote to offer a position to John Milnor, a professor at Princeton University. Oppenheimer objected that this violated the "gentlemen's agreement" between the two institutions not to poach from each other, and announced that he would forward his own recommendation to the board of trustees opposing the appointment. Selberg, as chairman of the School of Mathematics, furiously responded that the mathematicians would henceforth hold their meetings, which had previously taken place in Oppenheimer's office, without the director's presence at all. Gödel in turn was so upset by this defiance of proper authority that he stopped coming to faculty meetings altogether, voting by proxy instead after interminable phone discussions with his colleagues.[39]

When André Weil, Selberg's successor as chairman, announced his intention to continue the practice, Oppenheimer put his foot down—causing Weil to declare, "This means war!" Oppenheimer retaliated by establishing a separate committee for logic and mathematical foundations, consisting only of Gödel and Hassler Whitney, "who will meet with me as necessary," thereby removing at least one area of direct conflict with the rest of the mathematical faculty.[40]

The mathematicians were behind an even bigger uprising ten years later, which made the *New York Times* and was the subject of a tell-all cover story in *The Atlantic* ("Bad Days on Mount Olympus: The Shoot-Out at the Institute for Advanced Study"). When Oppenheimer's successor, the economist Carl Kaysen, announced his intention to appoint as a professor a sociologist of religion, Robert Bellah, the mathematicians gave scorching interviews denouncing not only Bellah's scholarly pretensions but Kaysen's moral and intellectual capacity and demanding his resignation. The Institute's physicists, possibly seeing a chance to get back at the trouble-making mathematicians, mostly supported Kaysen. Freeman Dyson archly told the *Times* that mathematicians regard religion as a "childhood disease." Gödel again was miserably caught in the middle. He agreed that Bellah was not qualified to hold a position at the Institute, but in the end was the only one among his mathematical colleagues not to vote against the nomination. (He abstained.) All of the internal politics deeply upset Gödel, who "was very distressed at the incivility of the atmosphere," a colleague recalled. As Gödel remarked to his mother a few years after his own appointment, "I often think with regret back to the lovely time when I did not yet have the honor of being a professor at the Institute."[41]

But along with all the trouble, his new position brought a much increased salary of $9,000 a year, which added greatly to his sense of security and to Adele's reconciliation to her new life. She had been far less happy with life in America in general, and Princeton in particular, than her husband, offering up a litany of complaints. "Adele unfortunately does not share my enthusiasm for this country at all," Gödel wrote to his mother six years after their arrival:

> She complains about everything and especially about the sanitary conditions. . . . Aside from that she hates life in a small town, and the fact that Trenton is only a ½ hour and New York only 1 hour from here is no compensation for her. The main reason for her discontent may well be that she is separated from her people with whom she spent all her time in Vienna. She finds it very difficult to connect with people here.[42]

Her difficulty learning English combined with her uninhibited unsophistication made Morgenstern fear a disaster from the start. "How is that ever supposed to work in Princeton society, namely with all the other normal wives," he wondered. Added to everything else were her own health problems, including appendicitis in 1946, the loss of most of her lower teeth, and her continual trouble keeping her weight under control, as well as ongoing psychological difficulties.[43] She overcompensated by dominating conversations and with occasional outbursts of wilder behavior. At a typically mad party at the Oppenheimers, a stand-up supper for a hundred guests, the evening ended with a very drunken Adele grabbing the young Freeman Dyson and dragging him onto the dance floor, forcing him to dance with her for twenty minutes while Gödel stood by himself in a corner looking miserable. "It makes me feel sick just to think of the horror of the lives these two people may be living," Dyson wrote in his diary that evening.[44]

But the 1950s brought a period of relative happiness and stability to both of them. As Gödel later told his psychiatrist Dr. Erlich, "I am not happily married and not unhappily married," but accepted his life as it was.[45] When Adele's second trip to Vienna, in 1953, gave his mother the opportunity of resuming her criticisms that Adele was squandering his money, he strongly rose to her defense, with objectivity and loyalty:

> I am not miserly with myself and would scarcely spend any more on myself if I had the most frugal wife in the world. . . . When you write that you now see that you have "always accurately judged Adele" and that she makes a great fuss and puts on an act, that is definitely false. Adele is by nature certainly entirely harmless and good-natured, but has clearly suffered a nervous breakdown made far worse by her experiences (especially the overly strict upbringing at home and her first marriage). That led especially to a dependence bordering on the pathologically exaggerated and an overvaluation of her family, and particularly of her mother (that was perhaps exploited by her people?) . . .
>
> In my last letter I did not lie you to at all. I did not keep from you that Adele has her sizeable flaws, only merely said that the resulting unpleasantnesses are more than balanced out by the good sides of my life here. And do you have anyone among your acquaintants who lives in a paradise and has nothing to complain about?[46]

Adele would continue to take periodic months-long vacations on her own—to the White Mountains of New Hampshire, the Jersey Shore, even Italy—which Gödel cheerfully accepted, no doubt recognizing that extended time apart helped ease the strains in their relationship.

In 1949 they had bought a house which Adele had fallen in love with. Morgenstern thought this another bit of Adele's foolishness: "Gödel has had a bad time. His wife is very hysterical, as he told me & described. He had to buy a house ($12,500 plus fees) that is worth much less and is not in a good area." The house, at 129 Linden Lane (since renumbered 145), was built right after the war, with a single story constructed of cinder block, lacking any attic ventilation or insulation. The Institute provided an advance on his salary for the $3,500 down payment, and the Gödels were both giddily pleased with their new possession.[47]

With all the pride of a new homeowner Gödel described to his mother "how lovely the place is": automatic oil furnace, wood-paneled living room, a fireplace with a shelf above to display their knickknacks. "It looks very cozy and I sometimes have the feeling I am in our villa. A garden plot is also included." Adele threw herself into the

With Adele at Linden Lane

work of gardening, planting bulbs and flowerbeds, shrubs and fruit trees, constructing with her own hands a small shade porch, and growing tomatoes and lettuce in the vegetable garden, all with an energy that reflected her newfound contentment with their now more settled existence, as well as being a source of wondering pride to her husband. "She had gotten it into her head to cut the grass every 8 days," he reported to Rudi. "Don't you think that is ridiculous?"[48]

And however much Adele's lowbrow tastes raised eyebrows among their sophisticated Princeton friends, in fact Gödel shared her distinctively Austrian weakness for the kitschy art, schmaltzy music, and sentimental knickknacks that filled their new house. Kreisel recalled that Gödel was proud of never changing his tastes, which he scorned as *Mangel an beständigen Gefühlen*, lack of consistency. He enthused in his letters to his mother over the cute music boxes, sentimental picture books, and little toy animals she sent at Easter and Christmas ("the ski bunny and two penguins are delightful"), the antics of the pet parakeets Adele kept, the animated Disney movies he never missed (*Snow White*, his favorite, he saw three times), and the Latin music and pop

Working outdoors

With the flamingo

songs on the radio and soap operas and other shows on daytime TV. "Many wrestling matches are broadcast, which of course do not interest me," he insisted, before adding, "There are by the way (you wouldn't think it possible) also wrestling matches between women and there are even a few quite pretty girls among them."[49]

He was especially proud of the concrete flamingo Adele bought for the garden, which she painted pink and black and placed in the middle of a flower bed right outside his study window. ("It really looks awfully cute," he told his mother, "especially when the sun shines on it.") His major amusement besides swimming, during two summers they spent vacationing at the Jersey Shore, was playing a Skee-Ball game on the boardwalk arcade; he proudly reported to his mother that he won a set of stainless steel kitchen utensils and an electric clock for Adele with his newfound skill.[50]

He and Adele did go see *Aïda* at the Metropolitan Opera one New Year's Day, and he later surprised Rudi with his knowledge and appreciation of modern art, and likewise never lost his interest in more serious literature—in his late fifties he read *Hamlet* ("The interesting thing is still that the various critics and literary historians are so divided in the interpretation of Hamlet's character") and a Gogol novella ("I was amazed how good it was"), along with German poetry and novels, and could discuss them knowledgeably and often with considerable psychological insight—but the gaps in his cultural knowledge popped up regularly.[51]

"Recently I discovered a modern writer previously unknown to me, 'Franz Kafka,'" he ingenuously informed his mother. "He writes quite insanely, but has a curiously vivid way of describing things." He had seen Wagner's operas *Lohengrin*, *The Flying Dutchman*, and *Tristan*, but had to ask his mother, "Who is 'Mahler'?" when he saw a reference to the birthday of the composer's widow, Alma. "Bach and Wagner," he confessed, "make me nervous."[52]

APPREHENSIONS OF EXISTENCE

In spite of his worries about the world situation and his ever-present anxieties over his health, Gödel, thought Hao Wang, was fundamentally a "rationalistic optimist." For most of the decade following his ulcer, his health was in fact as good as at any time in his life, and there were long periods of tranquility, order, and happiness in his life at home and work. He walked as before to the Institute every day with Einstein, "a half-hour, conversing all the while," arriving at his office about 11 a.m. and returning home around 2 p.m. to continue his reading or other work, sitting out in the garden on pleasant days. He from time to time would join his mathematical colleagues for lunch in town at the "stag room" of the Nassau Tavern and "seemed to enjoy these more intimate occasions," recalled his colleague Atle Selberg.[53]

At the same time, his hypochondria began undeniably to loom ominously larger. The etiology of hypochondria is still not well understood, but as with other anxiety disorders like obsessive-compulsive disorder it follows a self-reinforcing logic of its own, not necessarily tied to any definite external events. It is hard not to see at least a connection between his insistence on finding hidden causes in everyday affairs, his belief in the existence of absolute truth in his mathematical investigations, and his steadfast rejection of empirical medical evidence about his own condition. An intolerance of uncertainty is characteristic of most hypochondriacs, as is a tendency to perceive innocuous bodily symptoms as evidence of serious illness. The reassurance-seeking behavior that the anxiety provokes has the contradictory effect of only further reinforcing the underlying fears that drive it.[54]

Gödel began taking his temperature three times a day, and consulting a vast array of New York specialists in his conviction that he was suffering from any number of undiagnosed conditions, most of all the heart problems he was sure his doctors were overlooking (but which almost surely were garden-variety symptoms—such as an irregular heartbeat, which Gödel referred to as "palpitations"—of his own anxiety). A normal electrocardiogram only reinforced his belief that he knew more than his physicians, who would send him off with vague words of reassurance and prescriptions for vitamins or sleeping pills.[55] He and Adele continued to see psychiatrists from time to time, including the flamboyant German psychoanalyst Charles R. Hulbeck, who operated a posh practice from an office at 88 Central Park West, and who under his real name, Richard Huelsenbeck, had been a founder of the Dadaist art movement in Switzerland. Yet neither he nor Adele made a consistent commitment to addressing their psychological difficulties with life and their marriage.

Gödel's more serious problem was that his self-diagnoses were becoming self-fulfilling prophecies, as he began to dose himself with a staggering number of laxatives, antibiotics, and other drugs, all surely doing far more harm than good and inflaming or even creating the very problems he believed he was treating. He collected prescriptions and over-the-counter drugs and dispensed them according to his own theories: Gelusil and bicarbonate for indigestion (he had an alarm on his watch to remind himself when it was time for a dose); belladonna for irritable bowel; penicillin, aureomycin, or terramycin every time his temperature went up a fraction of a degree; and an entire medicine chest's worth of laxatives that he swallowed daily—milk of magnesia, lactose, Metamucil, Peri-Colase, ex-Lax, a patent elixir called Kurella.[56]

He had always paid excessive attention to his diet, but the exhaustive lists of prohibited foods that doctors in the 1950s prescribed to ulcer patients explains some of the more eccentric-sounding or infantile dietary choices he subsequently made—such as consuming a quarter of a pound of butter and five eggs a day, along with bland pureed baby foods and eggnog, all of which in fact reflected the "Bland Diet Instructions" he received from his physician. Yet at other times he still was able to enjoy Adele's baking and cooking, like the traditional Viennese crescent-shaped Christmas *Vanillekipferl* sugar cookies and a whole carp she prepared for Christmas Eve of 1958.[57]

His weight and mood swung together in an alternating series of rises and falls over the next two decades: "He is in very good shape," Morgenstern wrote in February 1954. "Imagines that he has severe heart problems & will soon die," a few months later. "Gödel is worse again." "He is completely out of his depression." "He looks good." "Looks miserable poor man." "In best mood and condition." "He has never been so thin . . . it is a miracle he is still alive." "He is better & I think everything will be all right again soon." "He looks like a living corpse." "A real miracle: he has gained 18 pd, is in excellent shape."[58]

Einstein's death in April 1955 brought intimations of mortality which Gödel characteristically responded to with evasion and denial. It was another night and day difference between the two friends. Years earlier Einstein had told his wife, "I have firmly decided to bite the dust with the minimum of medical assistance when my times comes, and up to then to sin to my wicked heart's desire." He vowed to eat whatever and sleep whenever he felt like, smoke on his pipe "like a chimney," and "go for a walk *only* in really pleasant company, and thus only rarely." Gödel was one of the rare few on whom that privilege was bestowed. Suffering a ruptured aortic aneurysm at age seventy-six, Einstein refused surgery, explaining, "I want to go when *I* want. It is tasteless to prolong life artificially."[59]

Gödel wrote his mother, "Of course I have lost much by his death, purely personally speaking, all the more since in particular lately he had become even nicer to me than he already was before, and I had the feeling that he wanted to come out of himself even more than before. . . .

After his death, I was asked twice to say something about him, but of course I declined."[60]

Einstein's death was the sudden end he had wanted, but von Neumann's death two years later from cancer was a prolonged ordeal which Gödel had a harder time facing. As the logician Gerald Sacks remarked with only slight exaggeration, "I noticed over the years that Gödel's way of cheering up a dying person was to send him a logical or mathematical puzzle." But it was also a manifestation of his own denial and fears of serious illness. In his last letter to von Neumann he expressed blithe confidence in his friend's imminent and complete recovery, before going on to pose what would be one of the most fundamental questions of computer science. Gödel's letter to his dying colleague was apparently the very first formulation of the so-called "P vs. NP" problem, which offered a striking analogy of his Incompleteness Theorem to the field of computing. "P" is the set of problems easy to solve, for example multiplication and addition. "NP" is the set of problems for which an efficient algorithm exists for *checking* a given solution, but finding the solution may or may not be easy, such as factoring a large number, solving a sudoku puzzle, or discovering a proof for a formula.[61]

Gödel pointed out that one could readily build a machine that works through every possible series of proof steps to discover whether a proof of n steps exists for a given formula. The crucial question is how rapidly the time required for the calculation increases as n gets larger. If it grows slowly—linearly, or even as a square of n—then in principle every problem that is easily checked ("NP") is also easily solved ("P"). If it grows exponentially, however, that means there will be a set of verifiable but, as a practical matter, forever uncomputable problems— just as his Incompleteness Theorem showed there are true but undecidable propositions within formal mathematical systems.

The answer remains open to this day. As the philosopher Wilfried Sieg observed, "In the face of human mortality, Gödel . . . chose to raise and discuss eternal mathematical questions."[62] He also showed that at age fifty he had not lost his genius for penetrating insight into fundamental matters of logic, when he still chose to exercise those gifts.

DREAMS AND NIGHTMARES OF VIENNA

In one of his first letters to his mother after the end of the war, Gödel expressed hope of returning to Vienna for a visit as soon as possible. But year after year he kept putting it off, making detailed plans and then canceling them, offering one excuse after another, of varying degrees of plausibility: the difficulty of getting travel permits, his uncertain health, the press of work, his special dietary needs, his and Adele's sudden decision to purchase the lot next to their house that left him short of money, his doctor's advice that he not travel in the summer, or stay anywhere lower than 1,000 meters elevation if he did.

But he finally admitted the real reason he never came. He was haunted by the events of 1936 and after: his mental breakdown and confinement in sanatoriums, the murder of Schlick, his anxiety over somehow being yet held responsible for Adele's abortion, and most of all his final, harrowing escape. "I am so happy to have escaped from lovely Europe that under no circumstances would I want to put myself at risk of not being able to come back here for any reason," he wrote. He subsequently admitted that even if that was not an entirely rational feeling, he could not shake it. "I was plagued for a time by nightmares that I went to Vienna and couldn't get back," he told his mother. "Now, nightmares are certainly no valid reason and so I planned to make the journey nonetheless, but the unpleasant feeling remains."[63]

In 1957, on the fifth and last time he reneged on his promised visit, he admitted that the nightmares had won out. "As to the things from 1936," he wrote his mother, "I spoke to a competent person recently and they were also not so absolutely sure if it is advisable for me to travel over there (especially now) (not because of anything that could happen in Austria)."[64]

Complicating his feelings was the series of shameful episodes in postwar Austria that showed how unwilling Austrians were to accept their responsibility for the events of the Anschluss and after that had uprooted the lives of Gödel and so many of his friends. The Austrian genius for looking the other way survived the war with flying colors.

After a brief and cursory period of denazification at the university, most of those responsible for the purges of Jews and liberals returned to the positions they had secured at the expense of their victims. Viktor Christian, the SS authority on the "Jewish Question" who had plundered Jewish libraries and graves, regained his university pension after a general amnesty by the government in 1950, and then was elected an honorary member of the Anthropological Society and, in 1960, honored with a Golden Doctorate from the University of Vienna on the fiftieth anniversary of earning his degree. Fritz Knoll, the Nazis' *kommissarische Rektor*, underwent a similar miraculous rehabilitation, elected general secretary of the Austrian Academy of Sciences in 1959 and bestowed with a special award from the university's academic senate in 1961 "in recognition of honorable and courageous leadership in difficult times." Arthur Marchet, who had been a member of the SA since 1933, leader of the Nazi Dozentenbund, and then named dean of the Faculty of Philosophy in 1943, had his degree suspended immediately after the war, only to have it restored in 1950. Within the Faculty of Philosophy, 77 percent of the professors were found to have been members of the Nazi Party, but two-thirds of those were able to regain academic positions in Austria within just a few years of the war's end, half of them to the very same posts they had held at the University of Vienna during the Nazi regime.[65]

Schlick's murderer, Hans Nelböck, survived the war, employed by the government as a petroleum engineer. In 1951 he sued Viktor Kraft for libel when the latter referred to him in the first published history of the Vienna Circle as "a paranoid psychopath." Fearing for good reason what Nelböck might do, Kraft agreed to an out of court settlement.[66]

Most shamefully of all, Leo Gabriel, the man who had goaded the impressionable Nelböck to his murderous deed not only walked scot-free, but in 1951 joined the ranks of the university's Faculty of Philosophy, where he remained for the next twenty years. Gabriel, like Nelböck, had been a destitute, far right-wing philosopher who hung around the university picking up crumbs of work. Only after Nelböck's trial did it come out that for months, Gabriel, who denounced Schlick as a "freemason," "Jew," "communist," and "Bolshevik" to anyone who would listen, had been feeding false stories to Nelböck that Schlick was the man responsible for his dismissal from his last paid job, a part-time teaching position at the Volkshochschule in Vienna. Gabriel conveniently absented himself during the trial, heading off to a "religious retreat" at a monastery in Innsbruck to avoid testifying. An utterly incompetent philosopher given to sententious and nebulous bloviations about "integral logic" and "synthesis of the whole," Gabriel rejected the scientific method altogether as corrupting and destructive. In the ultimate irony, the chair he was appointed to was the very one once held by Schlick, now rededicated to the study of Catholic philosophy.[67]

Almost no attempt was made to bring back those who had been fired or fled. The University of Vienna judiciously decreed in 1946 that because Menger had resigned his position in 1938, he could not "in a strict sense" count as having been dismissed by the Nazi regime.[68] As Gödel wrote his mother,

> That the Austrians today often do not want to give their colleagues abroad what is due them is probably true, and in part is motivated by material reasons. For actually (as the current Austrian government clearly views the Hitler regime as an illegal dictatorship), it ought to be obliged to rescind all firings from the universities. In most of the cases the victim would likely forego returning, but clearly the offer has not even been made them.[69]

Gödel subsequently turned down honorary membership in the Austrian Academy of Sciences, probably due more to his eccentric belief that since the academy had been established by the emperor he was forbidden as an American citizen from accepting the recognition than out of any desire to make a political statement. But his observations about postwar Austria always expressed an enduring ambivalence toward his homeland.

Finally realizing that Gödel was never going to make the trip back to Vienna, Marianne and Rudi came to Princeton in the spring of 1958 for a weeks-long visit. Despite whatever forebodings all might have had, the reunion of the family proved joyful and unstrained. The presence of

With Rudi and Marianne in Princeton

Adele's now ninety-year-old mother, who had been living with them in Princeton for the past two years after Adele decided she was not being properly looked after by her family in Vienna, might have added an extra element of tension to the visit, but Hildegard Porkert was a good-humored and uncomplaining woman "who does not trouble us in the least," Gödel had noted.[70] Gödel took his visitors to the Empire State Building and art museums in New York, and Marianne and Rudi thereafter repeated their visit to America every two years. "Adele and I were hugely happy about your visit and think it was a success in every regard," Gödel wrote on their return. "It is quite right that one is somewhat depressed after such a reunion. That is visible even in Adele who cries for you very much." Adele drew a picture at the bottom of the page depicting a heart, with tears pouring from it.[71]

· 10 ·

"If the World Is Constructed Rationally"

LIKE "A REALLY GOOD PLAY"

Since the 1940s there were persistent rumors in the mathematical world that Gödel had in his possession his long-sought proofs of the independence of the Continuum Hypothesis and Axiom of Choice—another stunning achievement, if true. Gödel told Morgenstern as much, confiding to him in 1942 that he was close to solving the first and a few years later that he had actually solved the second. The story was the talk of the 1950 International Congress of Mathematicians. But nothing from Gödel appeared. Martin Davis, a young logician who was in residence at the Institute for Advanced Study from 1955 to 1957, heard the rumors, too, and he and another young visitor, John Shepherdson, screwed up their courage to make an appointment with Gödel and ask if it was true. "I can't remember the details of the meeting," Davis said, "but it was awkward, and we came away without any information."[1]

In April 1963 Gödel received a letter from Paul J. Cohen, a twenty-nine-year-old American mathematician at Stanford University, outlining his discovery of the solutions that had eluded Gödel for a quarter century. Cohen said he would be in Princeton the following week to give a lecture and hoped to meet Gödel privately to explain his result.[2]

Unnerved by both the huge interest his discovery had aroused and the questions as to its validity that had been raised by prominent mathematicians, a panicked Cohen wrote Gödel a week later begging him to examine his manuscript thoroughly and sponsor it for publication in the *Proceedings of the National Academy of Sciences*, if it passed muster. "In short, what I am trying to say, is that I feel that only you, with your preeminent position in the field, can give the 'stamp of approval,' which I would so much desire. . . . If I have overstepped any bounds, please excuse me. I can only say that I feel under a great nervous strain . . ."[3]

Gödel dropped everything to examine the proof. "The verification of the accuracy of his solution was time-consuming," he wrote his mother, but his elation at finding that Cohen had indeed solved the problem, and in a startlingly original way, could scarcely have equaled what he would have felt had he achieved the result himself. It was a striking reflection of his great generosity of spirit and encouragement of others, even under circumstances that would have left many in his hypercompetitive field bitterly disappointed at having been beaten by a young upstart to a result he himself had labored on for years with success just out of reach. He never lost the deep aesthetic enjoyment of mathematics that he had absorbed in his heady youthful days in Vienna. "It is really a delight to read your proof of the independence of the continuum hypothesis," he began his reply to Cohen. "I think that in all essential respects you have given the best possible proof & this does not happen frequently. Reading your proof had a similar pleasant effect on me as seeing a really good play." Gödel followed up with another note urging Cohen not to worry about improving his result but to publish without delay, and to put aside his doubts and nervousness. "You have achieved the most important progress in set theory since its axiomatization," Gödel reassured him. "So you have every reason to be in high spirits."[4]

When the Princeton mathematician and philosopher Alonzo Church, in preparation for the presentation to Cohen of the Field Medal—the highest recognition in mathematics—asked Gödel a few years later to clarify what exactly he had accomplished toward a proof, Gödel replied forthrightly that the rumors of such had always been an exaggeration: "It is not true that, for many years, I have been in the possession of a proof of the independence of the continuum hypothesis and of the axiom of choice in set theory." He had, he explained, achieved in 1942 only a "partial result," but soon afterward, as a result of the shift of his interest to philosophy, "I never worked out this proof in detail and I never returned to these questions."[5]

Cohen, like most modern-day mathematicians, believed that his result meant it did not even make sense to speak of the Continuum Hypothesis or the Axiom of Choice (whose independence he was able to demonstrate using his same method) as being true or false; because they are neither contradicted nor implied by the other axioms of set theory, one can take them or leave them. Set theorists ever since have obtained interesting results from set theories with or without the Continuum Hypothesis. But that agnostic view never satisfied Gödel. He suspected that the Continuum Hypothesis would eventually be shown to be false, and he considered the Axiom of Choice to be obviously true. (The only time, Gerald Sacks said, he ever saw Gödel "sneer" was when the notion of rejecting the Axiom of Choice—as constructivist mathematicians did—came up. "I suppose it's interesting," Gödel drily replied, "to see what a man can do with one hand tied behind his back.")[6]

In the draft of a lecture he was invited to deliver to the American Philosophical Society in 1963 but in the end never gave, he expressed his sharpest disagreement with the entire worldview that "regards the world as an unordered and therefore meaningless heap of atoms." On the "left" stand skepticism, materialism, empiricism, and positivism—the values of Mach, the Vienna Circle, and most of modern science and philosophy. On the "right" are spiritualism, ideology, apriorism, and theology. Gödel made no bones about belonging to the "right," even as it placed him "in contradiction to the spirit of the time," indeed to the whole movement of philosophy and human knowledge since the Renaissance.[7] The human brain, he told Hao Wang, is "a computing machine connected with a spirit" that could not have come about through any mechanistic process. He occasionally astonished visiting members at the Institute by casually remarking that he did not believe in evolution or natural science. It was a divide over not only approach but attitude: skepticism and materialism implying a gloomy pessimism over human knowledge and even a grim nihilism toward life itself, Gödel argued; idealism and theology seeing "sense, purpose and reason in everything."[8]

While allowing some truth to both sides, Gödel passionately argued that rather than regarding even his Incompleteness Theorem as a pessimistic limitation on mathematical knowledge, the idealist viewpoint upholds the belief "that for clear questions posed by reason, reason can also find clear answers." It was true, he said, that if one regards the axioms from which mathematics starts as arbitrary choices that cannot be justified by any eternal or a priori principle, mathematics indeed does become a mere game played with symbols according to certain rules. But if the axioms represent a truth that the human mind can perceive as such, even though they cannot be formally derived from any rules or any process that a machine can imitate, then a way around such skeptical resignation is opened.[9] What he called "intuition" in this context had nothing to do with Brouwer's "intuitionism"; he was using the term to suggest that the human mind can literally see mathematical realities through a kind of perception, no different from the direct sensory perceptions that the empiricists decreed to be the only valid basis of physical laws. In an unpublished 1964 revision to "What Is Cantor's Continuum Problem?" he wrote,

> Despite their remoteness from sense experience, we do have something like a perception also of the objects of set theory, as is seen from the fact that the axioms force themselves upon us as being true. I don't see any reason why we should have less confidence in this kind of perception, i.e., in mathematical intuition, than in sense perception, which induced us to build up physical theories and to expect that future sense perceptions will agree with them, and, moreover, to believe that a question not decidable now has meaning and may be decided in the future. The set-theoretical paradoxes are hardly any more troublesome for mathematics than deceptions of the senses are for physics.[10]

In the formulation of axioms of mathematics, new axioms, which do not follow by any formal rule of logic, frequently become evident. And in the very fact that the undecidable propositions whose existence his Incompleteness Theorem implies include not just paradoxes of set theory or axioms like the Continuum Hypothesis—but real, substantive questions that all of our intuition tells us must have a simple yes–no answer, such as Goldbach's Conjecture—Gödel found a powerful reason to believe that our "intuitive grasping of ever newer axioms" will lead to their solutions. The astonishing fruitfulness of human intuition, already demonstrated in the fact that by imagining the axioms of set theory it is possible to derive properties of integers that can be verified by computation—and which might possibly also yield results verifiable in physics—testifies to the preexisting truth of the axioms so perceived, Gödel argued.[11]

Hilbert, he concluded, was mistaken on one point, namely in limiting the definition of mathematical truth to consistency of formal systems derived from axioms. The axioms too are part of mathematical truth, but of a kind that defies formalism altogether, accessible only via human intuition. "In any case," Gödel concluded, "there is no reason to trust blindly in the spirit of the time."[12]

His most personal views about God he kept almost completely to himself, sharing them only with his mother and only hinting at them to Morgenstern, fully aware of how out of keeping with the spirit of the time they were. In 1961, five years before Marianne's death at age eighty-six, he wrote her at length about his conviction that "other worlds," including an afterlife, exist.

You pose in your last letter the momentous question, whether I believe we shall meet in the hereafter. About that I can only say the following: If the world is constructed rationally and has a meaning, then that must be so. For what kind of a sense would there be in bringing forth a creature (man), who has such a broad field of possibilities of his own development and of relationships, and then not allow him to achieve 1/1000 of it. That would be approximately as if someone laid the foundation for a house with much effort and expenditure of money, then let everything go to ruin again. Does one have a reason to assume *that* the world is set up rationally? I believe so. For it is certainly not chaotic and arbitrary, but rather, as science shows, the greatest regularity and order reign in everything. . . . So, it follows directly that our earthly existence, since it in and of itself has at most a very dubious meaning, can only be a means to an end for another existence.[13]

While scorning the "nonsense" of most religious teaching, "e.g. according to Catholic dogma omnibenevolent God created most human beings exclusively for the purpose of sending them to Hell for all eternity," he noted that "even today's study of philosophy does not help much toward the understanding of such questions since after all 90% of today's philosophers see their primary task to be beating thoughts of religion out of peoples' heads, and thus having the same effect in this sense as the bad churches."[14]

In 1970, the year of his most severe psychological crisis since his dark year of 1936, he appeared at his office one day, full of life and bubbling conversation and told Morgenstern that he had completed both a proof that the value of the continuum was $\aleph_2$ and a logical proof for the existence of God, as Leibniz had attempted in his ontological proof. But later that year he was forced to withdraw the continuum paper, which he had sent to Tarski for submission to the *Proceedings of the National Academy of Sciences*, after a colleague, Robert Solvay, found that he simply could not make sense of the argument. "If it were anyone but Gödel," Solvay reported back to Tarski, "I would certainly recommend that the manuscript be rejected." In an unsent letter to Tarski, Gödel agreed it "is no good," explaining he had written it "in a hurry shortly after I had been ill, had been sleeping very poorly and had been taking drugs impairing the mental functions."[15]

As for his ontological proof—which he had been working on at the very same time he was falling into the grip of paranoid delusion at the start of the year—he told Morgenstern he hesitated to publish it because people would think "he really does believe in God," when in fact, he insisted, he had only carried out his proof as a "logical investigation" of what is possible with certain classical assumptions, such as "the absolute," appropriately axiomatized. Morgenstern jokingly told him he should publish it under a pseudonym, but that people would probably see through that—as had happened when Newton anonymously submitted his solution to the famous problem posed by Bernoulli of the curve of fastest descent of an object acted on by gravity, and Bernoulli had observed, "We recognize the lion by its claw."[16]

Gödel had in fact become rather famous by then in spite of himself, and not just in the small world of mathematical logic.

EVEN PRINCETON

Gödel's first inkling of public fame had come in the wake of the publication in *Scientific American* of a popular explanation of his Incompleteness Theorem in 1956, followed by a short book entitled simply *Gödel's Proof,* which became something of a mathematical cult classic. It almost did not happen. In his usual way, Gödel carried on an extended negotiation with the authors and publisher, Ernest Nagel and James R. Newman and New York University Press, over royalties and permission to reprint his original paper in the book, finally infuriating Nagel by demanding the right to see Nagel's complete manuscript and "eliminate" any passages he disapproved of. In the end, the book was published without Gödel's text or permission.[17] More attention came when Time-Life Books devoted several pages to him in a 1963 volume on mathematics, including a full-page portrait by the famous photographer Alfred Eisenstaedt.

A flood of mail arrived at the Institute from high school and college students, mathematicians from Scotland to South Africa, engineers at construction and aviation firms, the Department of Public Works in Baltimore, Maryland, and the U.S. Penitentiary in Terre Haute, Indiana, requesting his (long since unavailable) 1934 mimeographed lecture notes on the theorem.[18] Interview requests followed from the BBC, the *New York Times*'s science writer, and the Mathematical Association of America, which was producing a series of documentary films about living mathematicians, all of which he turned down. There was also a smaller deluge of fan and crank mail. An attractive young schoolteacher from Venice, California, sent a snapshot of herself posing in front of a large-format reproduction of the Eisenstaedt portrait posted on the wall of her classroom, which she explained in an accompanying note "serves two purposes: it intimidates the students, and it inspires me." An irrigation engineer in India and other self-taught amateurs sent him their supposed solutions to the four-color map problem, crackpot philosophical treatises, and (from an employee of an air-conditioning company) a proof that the Second Law of Thermodynamics implies the negation of the Axiom of Choice. Gerald Sacks thought Gödel was far too kind to all the cranks who would call up to talk to him. "He was an extremely courteous man," Sacks observed. "He wasted too much time I thought with crackpots."[19]

At the same time, Gödel's isolation from his professional colleagues was becoming a settled pattern. He never had students, and interacted only with the bravest of the visiting members in logic who dared to

Gödel's new office, in the modernist cube to the
right of the main library building

knock on his door. His office was now even more splendidly isolated, following his 1967 move to an annex off the new glass-and-concrete Social Sciences Library; his new workspace was at the rear of a small modernist tinted glass cube that stood by itself, looking out onto a pond and woods.

Still, the world was beginning to intrude in other ways in the 1960s. The student unrest that had been roiling the nation's campuses had largely bypassed sleepy conservative Princeton. The few students from the otherwise tranquil campus who had joined a march on Washington protesting the Vietnam War and draft in 1965 had carried a banner blazoned with the ironic slogan, "EVEN PRINCETON." ("They are right!" Gödel said of the student protesters at the time.)[20] But during Gödel's tormented spring of 1970, when paranoid delusions were adding a nightmarish dimension to his chronic anxiety and hypochondria, students and faculty members declared a four-day strike to denounce the expansion of the war into Cambodia. While Gödel was gripped with fantasies of dark specters breaking into his room and injecting him with drugs, and of his brother plotting behind his back, three hun-

Student protesters in Princeton, 1969

dred members of the radical Students for a Democratic Society were dragged off by police after they blocked the entrances and threw a brick through the window of the campus building where mathematicians and computer scientists of a university-run think tank, the Institute for Defense Analyses, worked on secret projects on codebreaking for the National Security Agency.[21]

When Gödel's psychiatrist Dr. Erlich moved his office at the end of the 1970 to another old Princeton dwelling a block away, he shared a hallway up a rattling wooden staircase in the quaint clapboard-sided house with the Fund for Peace Education and Draft Information, the Movement for a New Congress, and other antiwar and left-wing groups. The McGovern for President office would join them the following year. The anti-war groups, the local newspaper said, made 163 Nassau Street "one of the liveliest addresses in Princeton."[22]

Gödel continued to see Dr. Erlich once a week, his anxieties and obsessions now shifting to his wife's medical problems and his belief that *her* doctors were misdiagnosing and mistreating her worsening sciatica and polyneuropathy. But by the spring of 1971 he was once again on a much more even keel and ended his psychiatric appointments with Dr. Erlich, for the time being. Morgenstern found him again his old self: "Wonderful as always." "Great satisfaction from Gödel & our friendship." In the fall of 1973 Gödel even showed up for the director's big garden party to welcome the new members. "Gödel was very amusing. Holding 'court.' So many young logicians and their wives around him," Morgenstern recorded afterward.[23]

Even in the midst of the far worse blows to his physical and mental health that began in April 1974 with a sudden urinary blockage, he continued his pattern of alternating intervals of paranoid delusion and long stretches of cheerful lucidity. Morgenstern had been an unwilling witness to the onset of that medical crisis, which began with a protracted shouting match between Gödel and Adele over his refusal to accept the obvious: as his doctor immediately informed him, he was suffering from an enormously enlarged prostate. He finally agreed to go to the hospital but adamantly refused surgery, insisting at first that he could cure himself by taking milk of magnesia, then that he would

With Dorothy Morgenstern at the Institute garden party, 1973

just keep a catheter permanently in place so he could urinate. Morgenstern lamented: "How can so great a genius be so stubborn?!"[24]

The discomfort of a permanent catheter added to the misery of Gödel's last years; he was also dosing himself with a cocktail of antibiotics to ward off the ever-present risk of urinary infection. Morgenstern was again struck by the contrasts in his psyche: almost completely delusional regarding his doctors and health, yet full of "power, clarity, sharpness" when discussing mathematics and science. The day in September 1975 when Gödel received the National Medal of Science, he phoned Morgenstern "bubbling over" with excitement and pleasure at the honor and the heavy gold medal that accompanied it (he had refused to travel to Washington to attend the White House ceremony where President Gerald Ford conferred the awards, and the medal was sent to him afterward). Even as he lamented Gödel's obstinacy and increasingly emaciated appearance, Morgenstern frequently noted in his diary Gödel's undiminished flashes of genius and warm friendship.[25]

But Gödel was also becoming a trial, often phoning Morgenstern two or three times a day as Adele's deteriorating health left him ever more isolated and dependent: she had suffered a mild stroke in 1975 and was increasingly bedridden.[26] Morgenstern thought Gödel's attention-seeking complaints about his health and conspiratorial ideas about his doctors (he once told Morgenstern that the publicly available reference books listing medications were "false," and the doctors had "a secret one") were "in part play-acting"—Morgenstern's wife, Dorothy, had observed that there was always something "theatrical" about Gödel's behavior—but that did not make it any easier to take.[27]

Two days after Gödel's hospitalization in April 1974, Morgenstern had received a devastating diagnosis of his own. He learned he had metastasizing bone cancer, and feared that he would end up paralyzed, as had his friend John von Neumann two decades earlier. His love and patience toward Gödel added a tragic weight to his own painful end, marked by three years of surgeries and ending, just as he had dreaded, in paralysis.[28]

"A tragedy and how in all the world can I help him," Morgenstern wrote two weeks before his own death on July 26, 1977. "The doctors don't want him anymore, because he doesn't do at all what they want, etc. At the same time, believes that my paralysis will be over in a few days, and I will be up and new!" He wrote a memorandum for himself about his friend, full of heartbreak. "Here is one of the most brilliant men of our century, greatly attached to me . . . expecting help from me, and I unable to extend it to him. . . . By clinging to me—and he has nobody else, that is quite clear—he adds to the burden I am carrying." Gödel telephoned hoping to speak to Morgenstern on the day he died. When Dorothy Morgenstern told him the news of his last and closest friend's death, Gödel could manage to get out nothing but "Oh" before hanging up.[29]

That same month Adele was rushed to the hospital for emergency surgery that required a colostomy, then spent months in a nursing home recuperating. It was the final blow. Gödel's mental state had already deteriorated drastically over the past year, and at Adele's insistence he had started seeing Dr. Erlich again in February 1977. He was now insisting that Nazis were trying to have him declared mentally incompetent, the Internal Revenue Service was after him for failure to pay Social Security taxes for his hired help, the Institute was about to take away his pension, his brother had been sent to a concentration camp because Gödel had broken his promise to the Nazi government to return to Austria, he had made a fool of himself in America, people hated him, he was a terrible person, the doctors were impostors, evil spirits had removed from his desk a letter about his unfinished essay on Carnap, his friend Abraham Wald had actually not died in a plane crash nearly thirty years earlier but had survived and was secretly living in the Soviet Union. He had completely fallen out with his brother, apparently from festering resentment of Rudi's intervention in 1970 to have him get psychiatric help. "The fault is probably with Kurt," Morgenstern had sadly noted.[30]

Dr. Erlich had previously noted that whenever Gödel was ill, he underwent "emotional decompensation," a collapse of the psyche's usual coping mechanisms for dealing with stress. Now he was spiraling into an abyss of depression and paranoia. Erlich prescribed a high dose of a tricyclic antipsychotic drug, but it is not clear if Gödel ever took the medication. On August 28, 1977, Erlich came to his house to tell him in person that he had to go the hospital, but Gödel refused even when an ambulance Erlich had summoned arrived, and broke off his therapy sessions.[31]

The Gödels' neighbor Adeline Federici tried to help in Adele's absence, offering to go grocery shopping for him, but all Gödel ever wanted was Wonder bread, California navel oranges, and canned soup; when the price of soup went up two cents he refused to buy it anymore. To other visitors, including nursing help arranged for him, he refused even to open the door.[32]

The Institute administrators did what they could to reassure him that his wife's medical bills would be covered by his insurance and managed to arrange twenty-four-hour nursing care for Adele when, on December 19, she peremptorily discharged herself from the Princeton Nursing Home and had herself driven home by the Federicis. Until then, no one at the Institute had quite realized how terrible her husband's condition had become. He had spent the months of Adele's absence almost entirely alone and was slowly starving to death. Hao

Wang, who had been away from his position at Rockefeller University in New York City for most of the fall and had been unable to come to Princeton to visit Gödel, saw him for the last time two days before Adele's return. Wang was struck by the nimbleness of Gödel's mind, and thought he did not seem very ill. But Gödel told his friend sadly, "I have lost the faculty for making positive decisions. I can only make negative decisions."[33]

On December 29, Adele managed to persuade him to enter Princeton Hospital. He died there on January 14, 1978, curled in a fetal position. He had refused to eat at all during his final weeks. His death certificate listed the cause as "malnutrition and inanition secondary to personality disturbance."[34]

ROADS NOT TAKEN

A month and a half after her husband's death, and a day after receiving a statement from her husband's executor on her assets and future income, Adele entered Princeton's main jewelry store and picked out a large diamond ring and two bracelets costing $15,888. The owner was alarmed enough to contact the executor of her husband's estate, who informed the Institute's director, Harry Woolf. The lawyer handling the estate urged the Institute to arrange some financial counseling for Mrs. Gödel: "My strongest suggestion is that this be done and done soon."[35]

It was a warning of more trouble to come. Adele was left $115,000 in savings and an income totaling $30,000 a year from her husband's pension from the Institute plus Social Security and interest payments, but she required nursing and household help which she had trouble keeping due to her temperamental outbursts. At the end of the year she put the house and adjacent lot on Linden Lane up for sale and moved to a well-regarded assisted living center in nearby Doylestown, Pennsylvania, only to argue with the staff there and move on her own to a retirement community in Jamesburg, New Jersey, against her doctor's strong advice that she could not receive the medical care she needed in that setting. She died on February 4, 1981, the last years of her life as lonely and tragic as her husband's. In the end she had never really made peace with her new life in America.[36]

She left her husband's papers and literary rights to the Institute, though not before destroying all of the letters from Marianne and Rudi, despite Rudi's plea that they be preserved. She also apparently destroyed any letters between herself and her husband; not a single one has ever been found. For several years his boxes of papers lay piled in stacks six feet high in the basement of the Institute library, the bottom boxes starting to buckle under the weight of the ones above. John Dawson, a mathematics teacher at a community college in York, Pennsylvania, had been trying for several years, in connection with his interest in the history of symbolic logic, to find answers to the most basic questions about Gödel's life and publications, with almost no success. After repeatedly inquiring when Gödel's papers might be open to researchers, he received a call one day from the Institute suggesting that if he were serious about studying the papers, he should make a formal application to become a visiting member. The subsequent offer from the Institute added the proviso that, as long as he was at it, he should catalogue all of Gödel's papers.[37]

It took Dawson two years to go through the mountains of correspondence, unsent drafts of letters, receipts from plumbers, family photographs, and notebook after notebook filled with shorthand in incomprehensible Gabelsberger script. Eventually Dawson's wife, Cheryl, volunteered to learn Gabelsberger in order to decipher the more important items. But it was nearly forty years after Gödel's death before other researchers at last began transcribing the dozens of Gabelsberger notebooks containing his philosophical maxims and mathematical notes.

Gödel's public renown continued to grow after his death, notably boosted by the improbable 1979 bestseller *Gödel, Escher, Bach* by Douglas Hofstadter, which sought to weave Gödel's proof into an exploration of self-referential form in art, music, and ideas, but which contained not a word about the man himself. The general idea that there are truths that cannot be proved had an irresistible appeal that far outran Gödel's actual proof, helping to secure him a place, as the

American mathematician and writer Jordan Ellenberg put it, as "the romantic's favorite mathematician." Like Heisenberg's Uncertainty Principle and Einstein's Theory of Relativity, Gödel's Incompleteness Theorem has provided what Alan Sokal and Jean Bricmont, in their takedown of postmodernism, called "an inexhaustible source of intellectual abuse," invoked by theologians, literary theorists, architects, photographers, academic deconstructionists, pop philosophers, and mystics of all kinds to prove everything from the existence of God to the nature of free will, the structure of poetry, and the phenomenon of human misery. A not atypical example reads, "Basically, Gödel's theorems prove the Doctrine of Original Sin, the need for the sacrament of penance, and that there is a future eternity."[38]

Gödel himself, to be sure, was not always quite as scornful as his fellow logicians have since been of such interpretations. "It was to be expected after all that my proof would be made useful for religion sooner or later," he wrote his mother in 1963, "for that is doubtlessly supportable in a certain sense."[39] But probably more wrong things have been said about his proof than any other mathematical theorem in history, a dubious distinction—if perhaps an inevitable consequence of popular fame in a field such as mathematics.

Gödel's reputation and influence within science and philosophy has proved a more complex story. His relativity papers and philosophical insights have begun to receive considerably more respectful attention than they did from physicists and philosophers during his lifetime. When selecting the program for a memorial symposium at the Institute, the director scribbled "not worth a talk" next to "relativity" on a list of his major works, and his philosophical explorations were not even included on it.[40] Forty years later conferences and prestigious publication programs are being devoted to both topics.

The Incompleteness Theorem continues to cause trouble for both mathematics and philosophy, but in ways that are more subtle yet perhaps more intriguing than first imagined. What led his contemporaries to see a "catastrophe" for mathematics when the proof appeared undoubtedly seems less fearsome as time goes by. Avi Wigderson, one of Gödel's successors at the Institute for Advanced Study, observes that "mathematics has continued thriving." While Gödel certainly showed that some important mathematical propositions, notably the Continuum Hypothesis, are undecidable within certain formal systems, the question "Are there *interesting* mathematical truths that are *unknowable*" remains itself unanswered, as Wigderson points out. The self-referential statement Gödel constructed in his original proof was highly contrived—"unnatural," in the terminology of mathematicians—and since then only a few "natural" results in number theory or computing have been definitely shown to be undecidable in a Gödelean sense. In 1970, building on decades of work by Julia Robinson and others, Yuri Matisayevich proved that it is impossible to formulate a general rule for whether any given polynomial equation with integer coefficients, such as $3x^2 + 5y^3 + 2z = 0$, has a solution with integer values for all its unknowns—which was Hilbert's tenth challenge on his famous 1900 list of unsolved problems. In 1977, Jeff Paris and Leo Harrington established that a variant of Ramsey's Theorem, a problem in combinatorics, is undecidable within the system of Peano Arithmetic in which it can be formulated; and in 1982 Paris and Laurie Kirby did the same for Goodstein's Theorem, a similarly uncontrived proposition in number theory. But mostly, Gödel's proof as a practical matter has proved to be a barrier to a road that mathematicians have seen little reason to venture down.[41]

The fact that there might exist truths forever beyond human reach is likewise not necessarily the cause for philosophical alarm and despair it was taken to be by some. As the philosopher Robert Fogelin observed, skeptics about the extent of human knowledge come in two types, facetiously categorized as East Coast skeptics and West Coast skeptics according to their relative degree of laid-backness. "East Coast skeptics recognize that their knowledge is limited," Fogelin said, "and this troubles them deeply. West Coast skeptics recognize the same thing but find it liberating."[42]

Gödel himself firmly believed that his proof was profoundly encouraging for human creativity. Humans will always be able to recognize some truths through intuition, he consistently maintained, that can never be established even by the most advanced computing machine.

A machine that can literally duplicate the reasoning, learning, planning, and problem-solving ability of the human mind will be forever impossible if Gödel was right about what he believed to be the more far-reaching implications of his theorem. In place of limits on human knowledge and certainty, he saw only the irreplaceable uniqueness of the human spirit.

As Gerald Sacks keenly observed, "He made mathematics more interesting."[43] Though a "throwback as a philosopher," he never wavered on this: That any problem the human mind can pose, it can solve.

· Appendix ·
Gödel's Proof

GÖDEL CAME UP WITH several different versions of his scheme for encoding mathematical formulas as a single unique integer, but all begin by assigning a number to each of the "primitive" symbols used in the expressions of a formal mathematical system like Bertrand Russell's and Alfred North Whitehead's *Principia Mathematica*. In his 1934 lectures on the Incompleteness Theorem, Gödel assigned numbers 1 to 13 to the basic mathematical and logical symbols:

SYMBOL	DEFINITION	CODE NUMBER
0	zero	1
N	successor*	2
=	equals	3
~	not	4
∨	or	5
&	and	6
→	if . . . then	7
≡	if and only if	8
∀	universal quantifier	9
∃	existential quantifier	10
∈	member of a set	11
(	left parenthesis	12
)	right parenthesis	13

* The symbol $N(x)$ denotes the next greater number than x: thus $N(0)$ equals 1, etc.

Larger numbers designate variables and functions. For example, the variable x is assigned the number 16; y, 19; z, 22; and so on.

But as Gödel noted, the meaning of the symbols "is best forgotten" from this point on, just as a cipher clerk whose job it is to encode a message follows a rote procedure that makes no reference to the actual meaning of the words of the original message.

To ensure that a unique integer would be created for each formula, Gödel based his coding system on prime factors. To create the f-number for the formula "$0 = 0$" one takes the code number for each of its symbols:

$$
\begin{array}{ccc}
0 & = & 0 \\
\downarrow & \downarrow & \downarrow \\
1 & 3 & 1
\end{array}
$$

and uses them in turn as the exponents of the prime numbers (2, 3, 5, 7, 11, 13, . . .) arranged in sequence and multiplied together:

$$2^1 \times 3^3 \times 5^1 = 2 \times 27 \times 5 = 270$$

F-numbers become colossally large for ordinary mathematical expressions of any length or complexity; e.g., for the formula "$\sim \sim 0 = 0$"

$$
\begin{array}{ccccc}
\sim & \sim & 0 & = & 0 \\
\downarrow & \downarrow & \downarrow & \downarrow & \downarrow \\
4 & 4 & 1 & 3 & 1
\end{array}
$$

$$2^4 \times 3^4 \times 5^1 \times 7^3 \times 11^1 = 16 \times 81 \times 5 \times 343 \times 11,$$

the resulting f-number is 24,449,040. But Gödel's point was not that anyone would actually go to the trouble of making these arithmetic calculations, rather just that in principle every formula in a system like *Principia Mathematica* can be represented by a unique integer via this method; and moreover, the coding scheme permits every f-number to be "decoded" by reversing the process, breaking it down into its prime factors, then reading off the resulting exponents of each factor in sequence, to reveal the symbols that made up the original expression.

As Gödel pointed out, the f-numbers for any two formulas that follow one another according to the allowable rules of inference in a system will have some purely *numerical* relationship to each other. For example, the rule of inference which states that an expression of the form ~ ~ A (the negation of a negation) implies the expression A itself, corresponds to a numerical relation of the f-numbers of the pair of expressions that obey this rule, as in the example above, where "0 = 0" has an f-number that can be derived from the f-number of "~ ~ 0 = 0" by dropping the exponents of the first two factors ($2^4 \times 3^4$) then shifting the exponents that represent the remaining symbols of the formula two places to the left (leaving $2^1 \times 3^3 \times 5^1$).

To create a single integer representing an entire series of proof steps, Gödel repeated the process, using the f-number of each statement in the proof as the exponent of the series of prime numbers. So the f-number for the sequence of statements "~ ~ 0 = 0" and "0 = 0" would be given by $2^{270} \times 3^{24,449,040}$, an even more colossally large integer, but again one that in principle there is no reason could not be calculated. Every valid proof would begin with one or more of the axioms of the system, proceed by repeated applications of the rules of inference, and end with the statement to be proved. So the f-number for a valid proof would likewise embody numerical properties reflecting those conditions, which could be tested for by an appropriate (if immensely complex) arithmetic formula.

Gödel was thereby able to formulate a function *B*—itself a purely numerical expression that can be written as a formula in the symbols of *Principia Mathematica*—but which at the same time represents the metamathematical proposition, "The series of proof steps with f-number x is a proof of the formula with f-number z," written as,

$$x \; B \; z$$

The steps of Gödel's proof up to this point were ingenious, but his next step catapulted it into another realm entirely. The proposition,

"There is some value of x that satisfies the formula x B z"—which thus asserts that the formula with f-number z is provable—is, Gödel observed, similarly a formula that can be written within the system of *Principia Mathematica* using its defined symbols. He called this formula *Bew*, short for *Beweisbar*, "provable," defined by the expression,

$$Bew\ (z) = (\exists x)\ x\ B\ z$$

and its negation $\overline{Bew}$ as,

$$\overline{Bew}\ (z) = \sim (\exists x)\ x\ B\ z$$

which similarly asserts, "there is *no* value of x that satisfies the formula x B z": in other words, the formula with f-number z is *not* provable.

Having thus formulated a mathematical expression asserting that there is no series of possible steps that leads to the mathematical expression with formula number z, Gödel sought to find a way to turn that into a statement which refers to itself. To do so he needed to prove that there exists some number that, when substituted for z in this statement, results in an expression whose f-number is that substituted number itself—thus producing an expression declaring its own unprovability.

The trick he found to do this was a tour de force of self-reference. He began by defining a special function $Sub\ (y\ {}^{19}_{G\ (y)})$ that represents the f-number of the equation produced by the following process: start with the formula $G(y)$—defined as the formula with f-number y—and every time the variable with code number 19 appears within that formula, substitute for that variable the numerical expression that represents whatever numerical value y stands for.

He then inserted that function into $\overline{Bew}\ (z)$ to obtain,

$$\overline{Bew}\ (Sub\ (y\ {}^{19}_{G\ (y)}))\qquad\qquad formula\ A\ (\text{f-number } a).$$

Formula A possesses some astronomically large f-number, which of course Gödel did not attempt to calculate, but simply designated it as a.

He next wrote a new equation using the value of *a* in place of the variable *y*, and designated its f-number as *g*:

$$\overline{Bew}\ (Sub\ (a\ {}^{19}_{G\ (a)}))\qquad formula\ G\ (f\text{-number}\ g).$$

But 19 is the f-number for the variable *y* itself. So what $Sub\ (a\ {}^{19}_{G\ (a)})$ says is: go to the formula with f-number *a*—which is formula A—and then every time the variable *y* occurs, replace it with the expression for the number representing the value of *a*, and then calculate the f-number of the resulting formula. But that substitution of *a* for *y* is exactly the same alteration that transformed formula A to formula G in the first place: each occurrence of the variable *y* in formula A was replaced with the value of *a* to obtain formula G. So $Sub\ (a\ {}^{19}_{G\ (a)})$ equals none other than the f-value of formula G itself—namely *g*, yielding the desired result,

$$\overline{Bew}\ (g)\qquad formula\ G\ (f\text{-number}\ g),$$

the statement which asserts its own unprovability.

ACKNOWLEDGMENTS

Author's acknowledgments these days have a way of devolving into little more than self-congratulation and name-dropping. But I was helped by so many people who had no reason to do so other than their great generosity, shared interest in the life and work of Kurt Gödel, and deep commitment to confronting the history of Austria and Vienna in all its greatness and squalor, that I would be remiss in not at least trying to express here how deeply grateful and indebted to them I am.

First and foremost, Karl Sigmund, professor emeritus of mathematics at the University of Vienna and a walking encyclopedia of the intellectual history of his city, provided me with far more assistance than I had any right to ask, putting up with my interminable emailed questions and showing me around the sights of Gödel's Vienna, all with a generosity and sense of humor that did much to make the entire project a pleasure. His book on the Vienna Circle, *Exact Thinking in Demented Times*, was of enormous help to me in understanding the intellectual atmosphere of 1920s and 1930s Vienna, the scientific background of Gödel's own work, and the many fascinating personalities who moved in and out of his circle, as well as being a superb and immensely enjoyable piece of science writing and intellectual history. The published catalogue of the exhibition he organized in Vienna on the centenary of Gödel's birth in 2006, *Kurt Gödel: Das Album*, reproduces many key documents from his life and work and is another invaluable resource for all Gödeleans.

Marilya Veteto Reese, professor of German at Northern Arizona University, translated all of Gödel's letters to his family and went above and beyond the call of duty in many other ways, particularly by checking my translations of many other documents and letters. Her many discussions with me of the nuances of Austrian culture and language, and the wealth of her experience in the art of translation, did much to deepen my own understanding.

Christian Fleck, professor of sociology at the University of Graz, most generously offered to help me apply for research funding for the transcription of Gödel's extremely important 1937–38 diary notebook, written in the long-obsolete German system of Gabelsberger shorthand. With a grant from the Zukunftsfonds der Republik Österreich, we were able to commission an expert in Gabelsberger stenography, Dr. Erich Ruff, to carry out the transcription, with an English translation by Marilya Veteto Reese. (Complete texts may be found on my website, budiansky.com/goedel.) I would also like to express my deep thanks to Eva-Maria Engelen for providing me an early copy of her newly published transcription of another key Gödel shorthand notebook (*"Zeiteinteilung"*), a major contribution to the understanding of Gödel's inner life and thoughts.

Marcia Tucker, librarian of the Historical Studies–Social Science Library at the Institute for Advanced Study, offered nothing but encouragement and support from the outset. Besides formally acknowledging here the Institute's permission to quote from and reproduce material from the Kurt Gödel Papers, I would like especially to thank the Institute's archival specialist Erica Mosner and archival intern Max Lavery for searching out and scanning many documents from the Institute's own files. The staff of the Princeton University Library's Department of Special Collections also went to extraordinary lengths in the midst of the 2020 pandemic to provide me images from the Gödel Papers in time to meet my deadline, for which I am most appreciative.

Anita Eichinger, director of the Wienbibliothek im Rathaus, was instrumental in having all of Gödel's family letters in their collections digitized and made publicly available on their website. I am most grateful to her for so energetically and enthusiastically taking up my inquiry about whether this would be possible, and for her shared commitment to open access to archival sources.

NOTES

Full citations to sources cited in short form in the Notes below are found in the Bibliography. All of Gödel's correspondence with his family, unless otherwise noted, is from the Wienbibliothek im Rathaus, Vienna, the originals accessible on the Digital Wienbibliothek website. Frequently cited works and names are identified with the following abbreviations:

CW   Kurt Gödel. *Collected Works*. Edited by Solomon Feferman et al. 5 vols. New York: Oxford University Press, 1986–2003.

GA   Karl Sigmund, John Dawson, and Kurt Mühlberger. *Kurt Gödel: Das Album/The Album*. Wiesbaden: Vieweg, 2006.

IAS   Institute for Advanced Study. Archives. Files on Kurt Gödel. Princeton, N.J.

KG   Kurt Gödel

KGP   Kurt Gödel Papers. Institute for Advanced Study Archives. On deposit at the Manuscripts Division, Special Collections, Princeton University Library, Princeton, N.J.

MG   Marianne Gödel (mother of KG)

OH   Oral History

OMD   Oskar Morgenstern. Diaries. Oskar Morgenstern Papers. Rare Book and Manuscript Library, Duke University, Durham, N.C. Digital publication on the Web, "Oskar Morgenstern Tagebuchedition," University of Graz, Austria.

RG   Rudolf ("Rudi") Gödel (brother of KG)

W&B   Eckehart Köhler et al., eds. *Wahrheit & Beweisbarkeit*. Vol. 1, *Dokumente und historische Analysen*. Vienna: öbv & hpt, 2002.

PROLOGUE

1. The direct and indirect quotations from the notes of KG's psychiatrist here and below are from Dr. Philip Erlich File on Kurt Gödel, KGP, 27/1.
2. OMD, 23 September 1974; KG's height is given in Certificate of Naturalization, 2 April 1948, KGP, 13a/6 (Series 8). A recording he made of a birthday greeting to his mother in 1949 can be heard in the DVD documentary, DePauli-Schimanovich and Weibel, *Ein mathematischer Mythos.*
3. Albert Einstein quoted in Oskar Morgenstern to Bruno Kreisky, 25 October 1965, reprinted in *W&B*, 23–24.
4. KG to MG, 1 November 1950.
5. Carl Kaysen to KG, 13 April 1970, IAS, Faculty Files, 1951–1977.
6. Dr. Harvey Rothberg to Apostolos Doxiadis, email, 4 July 2001, KGP, 27/1.

CHAPTER 1: DREAMS OF AN EMPIRE

1. Rudolf Gödel, "Chronik der Familie," 55.
2. Roth, *Emperor's Tomb*, 26; Johnston, *Austrian Mind*, 45; Beller, *Vienna and the Jews*, 176.
3. Roth, *Emperor's Tomb*, 28.
4. Berkley, *Vienna and Its Jews*, 367.
5. Roth, *Emperor's Tomb*, 28.
6. Zweig, *World of Yesterday*, 23–24.
7. Janik and Toulmin, *Wittgenstein's Vienna*, 42.
8. Hofmannsthal, "Die österreichische Idee," 405; Roth, *Emperor's Tomb*, 114.
9. Spiel, *Vienna's Golden Autumn*, 26.
10. Judson, *Habsburg Empire*, 63–66; Beller, *History of Austria*, 98–99.
11. Beller, *History of Austria*, 100; Judson, *Habsburg Empire*, 61–62.
12. Judson, *Habsburg Empire*, 62, 65–66; Deak, "Austrian Civil Service," 64, 135–36; Heindl, *Gehorsame Rebellen*, 102.
13. Hamann, *Hitler's Vienna*, 6.
14. Beller, *History of Austria*, 98–100; Judson, *Habsburg Empire*, 61–63.
15. Winkler, "Population of Austrian Republic," 1.
16. Spiel, *Vienna's Golden Autumn*, 34.
17. Judson, *Habsburg Empire*, 156.
18. Spiel, *Vienna's Golden Autumn*, 31; Janik and Toulmin, *Wittgenstein's Vienna*, 38; Beller, *History of Austria*, 116–17.

19. May, *Hapsburg Monarchy*, 145; Johnston, *Austrian Mind*, 33–34.

20. Beller, *History of Austria*, 131; Deak, "Austrian Civil Service," 247.

21. Musil, *Mann ohne Eigenschaften*, 31.

22. Johnston, *Austrian Mind*, 50–51.

23. Roth, *Radetzky March*, 192–93.

24. Hamann, *Hitler's Vienna*, 90.

25. Schorske, *Fin-de-Siècle Vienna*, 25–27.

26. Beller, *History of Austria*, 167; Spiel, *Vienna's Golden Autumn*, 41.

27. Germaine de Staël quoted in Spiel, *Vienna's Golden Autumn*, 38.

28. Johnston, *Austrian Mind*, 224, 226–27.

29. Von Kármán, *Wind and Beyond*, 21; Smith, "Elusive Dr. Szilard."

30. Johnston, *Austrian Mind*, 132.

31. Beller, *Vienna and the Jews*, 34, 51, 166.

32. Aly, *Why the Germans*, 15.

33. Aly, *Why the Germans*, 23–25; Beller, *Vienna and the Jews*, 89, 92.

34. Sigmund Freud quoted in Beller, *Vienna and the Jews*, 208.

35. Beller, *Vienna and the Jews*, 39; Karl Kraus quoted in Spiel, *Vienna's Golden Autumn*, 39–40.

36. Beller, *Vienna and the Jews*, 90 and n. 12; Monk, *Wittgenstein*, 228.

37. Graf, *Musical City*, 65.

38. Janik and Toulmin, *Wittgenstein's Vienna*, 34, 50–51; Hamann, *Hitler's Vienna*, 147–49.

39. Zweig, *World of Yesterday*, 104–5, 110.

40. Spiel, *Vienna's Golden Autumn*, 32; Klemens von Klemperer quoted in Berkley, *Vienna and Its Jews*, 22.

41. Musil, *Mann ohne Eigenschaften*, 34; Wickham Steed quoted in Beller, *Vienna and the Jews*, 175; Johnston, *Austrian Mind*, 48.

42. Viktor Adler quoted in Johnston, *Austrian Mind*, 22–23.

43. Bahr, *Wien*, 72; Franz I quoted in Spiel, *Vienna's Golden Autumn*, 33.

44. Johnston, *Austrian Mind*, 50.

45. Hamann, *Hitler's Vienna*, 284–90; for "Judapest" see Connolly, "Lueger," 248 n. 3.

46. Hamann, *Hitler's Vienna*, 285–86, 290.

47. Zweig, *World of Yesterday*, 85–86.

48. Hermann Broch quoted in Beller, *Vienna and the Jews*, 176.

49. Schorske, *Fin-de-Siècle Vienna*, 10, 304.

50. Johnson, *Introducing Austria*, 171; Anton Kuh quoted in Spiel, *Vienna's Golden Autumn*, 57.

51. Musil, *Mann ohne Eigenschaften*, 35 (my translation).

52. Sigmund, *Exact Thinking*, 13, 20–23; Spiel, *Vienna's Golden Autumn*, 134.

53. Roth, *Emperor's Tomb*, 91; Hermann Broch quoted in Beller, *Vienna and the Jews*, 177.
54. Kraus, "Franz Ferdinand," 2; Sigmund Freud quoted in Johnston, *Austrian Mind*, 238.

CHAPTER 2: *ALLE ECHTEN WIENER SIND AUS BRÜNN*

1. Berend, *European Economy*, 192.
2. List, *Deutsches Bollwerk*.
3. Quotations of Rudi Gödel and accounts of the Gödel family here and below are from Rudolf Gödel, "Chronik der Familie," 51–57, unless otherwise noted.
4. KG to MG, 27 June and 31 July 1954; Johnston, *Austrian Mind*, 267.
5. Dr. Philip Erlich File on Kurt Gödel, 2 February 1971, KGP, 27/1.
6. Engelen, ed., *Notizbücher*, 395; KG to MG, 26 August 1946.
7. KG to MG, 14 August 1961; KG to RG, 31 January 1952.
8. KG to MG, 25 June 1961.
9. Rudolf Gödel, "History of Gödel Family," 15.
10. KG to MG, 30 April 1957.
11. KG to MG, 27 February 1950.
12. Notebook entry, n.d., quoted in *CW*, 4:425.
13. Rothkirchen, *Jews of Bohemia and Moravia*, 10–11.
14. Schober-Bendixen, *Tuch-Redlichs*, 32–36, 127.
15. Rothschild, *Europe Between Wars*, 79–80; Judson, *Habsburg Empire*, 434–35.
16. Judson, *Habsburg Empire*, 314–16, 377.
17. Rothschild, *Europe Between Wars*, 80–81, 81 n. 4.
18. Zahra, *Kidnapped Souls*, 106, 121–22; Rothschild, *Europe Between Wars*, 111.
19. Judson, *Habsburg Empire*, 444–45.
20. Rothschild, *Europe Between Wars*, 86–87.
21. Johnson, *Introducing Austria*, 63–66; Beller, *History of Austria*, 203.
22. Zweig, *World of Yesterday*, 305, 306–7.
23. Slezak, *Rückfall*, 177.
24. KG mentions his appendectomy in KG to MG, 14 February 1962.
25. "Frequentations-Zeugnis," 5 July 1916, KGP, 13a/0.
26. School reports, Staats-Realgymnasium, KGP, 13a/0.
27. KG to MG, 31 July 1954; Harry Klepetar quoted in Dawson, *Logical Dilemmas*, 15.
28. Klepetar, "Chances of Communism," 301n.

29. "Deaths," *Journal of the American Medical Association* 168, no. 10 (November 8, 1958): 1388; KG to RG, 25 May 1941; Claims Resolution Tribunal, Adolf Hochwald, Case No. CV96-4849, 21 April 2003.

30. KG to MG, 16 April 1949.

31. Harry Klepetar quoted in Dawson, *Logical Dilemmas*, 17; KG to MG, 11 September 1960.

32. KG to Burke D. Grandjean, unsent reply to questionnaire ca. 1976, *CW*, 4:446–47.

CHAPTER 3: VIENNA 1924

1. Sigmund Freud quoted in Johnston, *Austrian Mind*, 238.

2. Kraus, *Last Days*, 3; Patterson and Pyle, "1918 Influenza Pandemic," 14.

3. Clare, *Last Waltz in Vienna*, 39; Max von Laue to Moritz Schlick, 3 September 1922, quoted in Sigmund, *Exact Thinking*, 105–6.

4. Menger, *Reminiscences*, 2.

5. Menger, *Reminiscences*, 2; Beller, *History of Austria*, 206–8.

6. Taussky-Todd, OH, 14; Kuh, "Central und Herrenhof," 22; Segel, ed., *Coffeehouse Wits*, 27–28.

7. Ludwig Koessler to Albert Einstein, 31 December 1920, Einstein, *Collected Papers*, 12 (English trans. supp.): 422 and 35–36 n. 5.

8. Sigmund, "Hans Hahn," 27; Karl Menger interview, *W&B*, 227; Menger, *Reminiscences*, 9.

9. Hamann, *Hitler's Vienna*, 216–27; Menger, *Reminiscences*, 14–16.

10. Hilbert, "Mathematische Probleme," 262.

11. Taussky-Todd, OH, 10–11.

12. Taussky-Todd, OH, 1–10.

13. Taussky-Todd, "Remembrances of Gödel," 35–36.

14. List of KG's enrolled courses, *W&B*, 145–46.

15. Kreisel, "Kurt Gödel," 153; Wang, *Reflections on Gödel*, 18.

16. Heinrich Gomperz quoted in Stadler, *Vienna Circle*, 235.

17. Taussky-Todd, OH, 17–18; *GA*, 23.

18. KG, "Protokolle," 10.

19. Menger, *Reminiscences*, 212–13 (economic theory); OMD, 13 January 1948 (fog statistics); Menger, *Reminiscences*, 71–72 (Hegel); OMD, 23 January 1947 (*World Almanac*).

20. Gödel preserved his library request slips: see Wang, *Reflections on Gödel*, 21; Dawson, *Logical Dilemmas*, 25.

21. Sigmund, *Exact Thinking*, 7.

22. Hans Hahn to Paul Ehrenfest, 26 December 1909, quoted in Sigmund, *Exact Thinking*, 79.

23. Sigmund, "Hans Hahn," 18; Menger, "Introduction," ix.

24. Menger, "Introduction," ix; Hans Hahn quoted in Sigmund, "Hans Hahn," 25–26.

25. List of KG's enrolled courses, *W&B*, 145; Hans Hahn quoted in Menger, "Introduction," xi.

26. Russell, *Autobiography, 1914–1944*, 29–30.

27. Menger, *Reminiscences*, 57; Karl Popper quoted in Sigmund, "Hans Hahn," 17.

28. Menger, "Introduction," xi, xvii.

29. Menger, *Reminiscences*, 201; Wang, *Reflections on Gödel*, 41; OMD, 12 May 1949.

30. Engelen, ed., *Notizbücher*, 346–47; OMD, 12 May 1949; KG to MG, 21 April 1965.

31. Broch, *Unknown Quantity*, 42.

32. Broch, *Unknown Quantity*, 27.

33. This famous anecdote is quoted with the most convincing source of its authority in Adani, "Hans Bethe," 96–97.

34. Hilbert, "Mathematical Problems."

35. Richard Courant quoted in Reid, *Hilbert*, 111.

36. Reid, *Hilbert*, 109, 131–32.

37. Weyl, "Hilbert," 612.

38. Curbera, *Mathematicians of the World*, 85; Reid, *Hilbert*, 188.

39. Siegmund-Schultze, "Mathematics Knows No Races," 58–59; Fraenkel, *Recollections*, 137.

40. Curbera, *Mathematicians of the World*, 88–89.

41. David Hilbert quoted in Reid, *Hilbert*, 185, and Sigmund, "Hans Hahn," 23.

42. Sigmund, *Exact Thinking*, 165; Menger, *Reminiscences*, 5; Stadler, *Vienna Circle*, 285.

43. Menger, *Reminiscences*, 5; Stadler, *Vienna Circle*, 298–99.

44. Quoted in Beller, *History of Austria*, 212.

45. Pauley, *Forgotten Nazis*, 17–18.

46. Aly, *Why the Germans*, 73–74, 129–30, 173; Weber, "Science as Vocation," 134.

47. Sigmund, *Exact Thinking*, 173; Pauley, *Prejudice to Persecution*, 121–22; Stadler, *Vienna Circle*, 291–92.

48. "Terror against the Anatomical Institute of Julius Tandler," University of Vienna, *650 Plus*.

49. Stadler, *Vienna Circle*, 289; Sigmund, *Exact Thinking*, 173.

50. "The 'Bärenhöhle,'" University of Vienna, *650 Plus*.

51. Stadler, *Vienna Circle*, 293–95; Sigmund, *Exact Thinking*, 175.

52. Beller, *History of Austria*, 210–12; Johnson, *Introducing Austria*, 99–101.

53. Marcel Natkin to KG, 20 July 1927, KGP, 2c/114.

54. Clare, *Last Waltz in Vienna*, 116.

CHAPTER 4: FLOATING IN MIDAIR

1. Sigmund, *Exact Thinking*, 108–9, 244; Menger, *Reminiscences*, 54–55.

2. Sigmund, *Exact Thinking*, 108; Feigl, "Wiener Kreis," 631.

3. "Olga Hahn-Neurath," University of Vienna, *650 Plus*; Robert Musil quoted in Sigmund, *Exact Thinking*, 89.

4. Menger, *Reminiscences*, 16–17, 69; Stadler, *Vienna Circle*, 54.

5. Menger, *Reminiscences*, 54.

6. Menger, *Reminiscences*, 55; Moritz Schlick to Albert Einstein, 12 June 1926, Schlick, Einstein Correspondence.

7. Menger, *Reminiscences*, 224–25.

8. Arthur Schnitzler quoted in Sigmund, *Exact Thinking*, 202–4.

9. Menger, *Reminiscences*, 38–43.

10. Menger, *Reminiscences*, 200.

11. KG to MG, 15 August 1946.

12. Sigmund, *Exact Thinking*, 116–18; Feigl, "Wiener Kreis," 636 ("engineer"), 655 ("verbal sedatives").

13. Stadler, *Vienna Circle*, 53.

14. Feigl, "Wiener Kreis," 630.

15. Menger, *Reminiscences*, 70.

16. Carl Gustav Hempel interview, *W&B*, 262.

17. Hans Hahn quoted in Sigmund, *Exact Thinking*, v.

18. Sigmund, *Exact Thinking*, 5, 178.

19. Rudolf Carnap quoted in Sigmund, *Exact Thinking*, 165; Carnap diary, 10 September 1931, quoted in Wang, *Reflections on Gödel*, 91.

20. Menger, "Introduction," xviii n. 11.

21. Feigl, "Wiener Kreis," 634.

22. Waugh, *House of Wittgenstein*, 29–30, 34.

23. Bertrand Russell to Ottoline Morrell, 1 and 2 November 1911, quoted in Monk, *Wittgenstein*, 39; Russell, *Autobiography, 1914–1944*, 137.

24. Russell, *Autobiography, 1914–1944*, 136–37.

25. Ludwig Wittgenstein to G. E. Moore, 7 May 1914, quoted in Monk, *Wittgenstein*, 103.

26. Monk, *Wittgenstein*, 102.

27. Monk, *Wittgenstein*, 138; Ludwig Wittgenstein to Bertrand Russell, 13 March 1919, Russell, *Autobiography, 1914–1944*, 162.

28. Russell, *My Philosophical Development*, 88; Feigl, "Wiener Kreis," 634, 637.

29. Moritz Schlick to Albert Einstein, 14 July 1927, Schlick, Einstein Correspondence.

30. Sigmund, *Exact Thinking*, 127, 133–34; Waugh, *House of Wittgenstein*, 148–49.

31. Monk, *Wittgenstein*, 242–43.

32. Feigl, "Wiener Kreis," 639; Menger, *Reminiscences*, 131. In a letter to Karl Menger in 1972, Gödel recalled seeing Wittgenstein the one and only time in his life "when he attended a lecture in Vienna. I think it was Brouwer's": KG to Menger, 20 April 1972, *CW*, 5:133.

33. KG to Burke D. Grandjean, unsent reply to questionnaire, ca. 1976, *CW*, 4:447, 450. For doubts about the consistency of Gödel's Platonist views, see Feferman, "Provably Unprovable"; Davis, "What Did Gödel Believe"; Martin, "Gödel's Conceptual Realism."

34. Gespräch mit Gödel über Logik und Mathematik, Rudolf Carnap Diary, 14 December 1928, Carnap Papers, Gödel Allgemein, 102b/43.

35. Dr. Philip Erlich File on Gödel, 10 August 1970, KGP, 27/1; KG, "Is Mathematics Syntax of Language?"

36. KG to Burke D. Grandjean, January 1976, *CW*, 4:443–44.

37. Marcel Natkin to KG, 20 July 1927, KGP, 2c/114.

38. Menger, *Reminiscences*, 210.

39. Menger, *Reminiscences*, 201.

40. Taussky-Todd, "Remembrances of Gödel," 33, 36.

41. KG, "Protokolle," 10–11.

42. Menger, *Reminiscences*, 205.

43. Karl Sigmund, personal communication; Feigl, "Wiener Kreis," 640; Herbert Feigl to KG, 22 November 1956 and 21 February 1957, KGP, 1c/46.

44. OMD, 27 May 1958, 9 September 1974, and 25 October 1969.

45. Rucker, *Infinity and Mind*, 178.

46. Sacks, "Reflections on Gödel."

47. Taussky-Todd, "Remembrances of Gödel," 32–33.

48. Adele's listing for massage services appears in *Lehmann*, the Vienna city directory, for the years 1929 to 1931. Although the frequently mentioned account of Adele's having been a nightclub dancer has been challenged (notably by DePauli-Schimanovich, *Gödel und Logik*, 405), Gödel him-

self stated it as a fact to his psychiatrist Dr. Erlich in Princeton in 1970 (Dr. Philip Erlich File on Kurt Gödel, 21 May 1970, KGP, 27/1).

49. Whitehead and Russell, *Principia Mathematica*, 2:86.
50. Russell, *Autobiography, 1872–1914*, 221–22; Bertrand Russell to Jean van Heijenoort, 23 November 1962, in Heijenoort, ed., *Frege to Gödel*, 127.
51. Frege, *Grundgesetze der Arithmetik*, 2:253.
52. Musil, "Der mathematische Mensch."
53. KG, "On Undecidable Propositions," *CW*, 1:362; Weyl, "Mathematics and Logic," 6.
54. Russell, *Autobiography, 1872–1914*, 229; Monk, *Russell*, 193.
55. Sales receipt, Franz Deuticke Buchhandlung, 21 July 1928, reproduced in *GA*, 110; Kurt Gödel to Herbert Feigl, 24 September 1928, *CW*, 4:403.
56. Snapper, "Three Crises in Mathematics," 208–9.
57. David Hilbert quoted in Hoffmann, *Gödel'schen*, 35–36; Hilbert, "On the Infinite," 375.
58. Fogelin, "Wittgenstein and Intuitionism," 273.
59. Hans Hahn quoted in Sigmund, "Hans Hahn," 26, and Sigmund, *Exact Thinking*, 151–52. I have slightly adjusted the quoted translation from the original German.
60. Sigmund, *Exact Thinking*, 213.
61. Fogelin, "Wittgenstein and Intuitionism," 268–69.
62. Russell, *Mathematical Philosophy*, 179.
63. KG, "Situation in Foundations of Mathematics," *CW*, 3:49.
64. David Hilbert quoted in Hoffmann, *Gödel'schen*, 35.
65. Hilbert quoted in Reid, *Hilbert*, 157, 184.
66. Brouwer, "Reflections on Formalism," 492; *CW*, 1:49; Hoffmann, *Gödel'schen*, 27–28.
67. KG, "Lecture at Zilsel's," *CW*, 3:93.
68. KG, "Existence of Undecidable Propositions," 3.

CHAPTER 5: UNDECIDABLE TRUTHS

1. KG, "Vollständigkeit des Logikkalküls"; Wang, "Facts about Gödel," 654 n. 2.
2. KG, "Vollständigkeit der Axiome."
3. Sacks, "Reflections on Gödel."
4. Rudolf Gödel, "Chronik der Familie," 58.

5. Herbert Feigl to KG, 29 March 1929, KGP, 1c/45.

6. KG, "Protokolle," 10; KG to MG, 30 September 1956.

7. Rudolf Gödel, "Chronik der Familie," 58.

8. KG income/expense ledger, KGP, 13b/31.

9. Stadler, *Vienna Circle*, 153–55.

10. Program of the Second Conference on the Epistemology of the Exact Sciences, Königsberg, September 5–7, 1930, reprinted in Stadler, *Vienna Circle*, 162–63.

11. Sigmund, *Exact Thinking*, 221.

12. Von Plato, "Sources of Incompleteness," 4047.

13. Rudolf Carnap diary quoted in Wang, *Reflections on Gödel*, 84.

14. Rudolf Carnap diary quoted in Wang, *Reflections on Gödel*, 85.

15. KG, "Lecture at Königsberg," *CW*, 3:28.

16. Von Plato, "Sources of Incompleteness," 4047; Hahn et al., "Diskussion zur Grundlegung," 148.

17. Hahn et al., "Diskussion zur Grundlegung," 148.

18. KG's postscript in Hahn et al., "Diskussion zur Grundlegung," 147–51; KG, "Über unentscheidbare Sätze."

19. Rucker, *Infinity and Mind*, 182; Engelen, ed., Notizbücher, 376, 390.

20. Enzensberger, "Hommage à Gödel," reprinted in *W&B*, 25 (my translation).

21. KG, "Existence of Undecidable Propositions," 6.

22. KG, "Existence of Undecidable Propositions," 6–7.

23. KG, "Situation in Foundations of Mathematics," *CW*, 3:50–51.

24. KG, "Existence of Undecidable Propositions," 8–9.

25. KG, "Undecidable Propositions of Formal Systems," *CW*, 1:355.

26. KG, "Existence of Undecidable Propositions," 14.

27. KG, "Undecidable Propositions of Formal Systems," *CW*, 1:359.

28. Kleene, "Kurt Gödel," 154.

29. Vinnikov, "Hilbert's Apology."

30. Heinrich Scholz to Rudolf Carnap, 16 April 1931, quoted in Mancosu, "Reception of Gödel's Theorem," 33; Marcel Natkin to KG, 27 June 1931, KGP, 2c/114.

31. Goldstine, *Pascal to von Neumann*, 167–68.

32. John von Neumann to KG, 20 November 1930, *CW*, 5:336–39.

33. Drafts of KG to John von Neumann, late November 1930, quoted in von Plato, "Sources of Incompleteness," 4050–51.

34. John von Neumann to KG, 29 November 1930, *CW*, 5:338.

35. Von Plato, "Sources of Incompleteness," 4054.

36. Ulam, *Adventures of a Mathematician*, 80; Goldstine, *Pascal to von Neumann*, 174.

37. Statement in Connection with the First Presentation of the Albert Einstein Award to Dr. K. Gödel, March 14, 1951, Einstein, Faculty Files; Gustav Hempel interview, *W&B*, 253–54.

38. Afterword by Franz Alt in Menger, *Ergebnisse*, 469–70.

39. Menger, *Reminiscences*, 202–4.

40. Reed, *Hilbert*, 198.

41. David Hilbert quoted in Dawson, "Reception of Gödel's Theorem," 267 n. 10.

42. Taussky-Todd, OH, 16.

43. Ernst Zermelo to KG, 21 September 1931 and 29 October 1931, KG to Zermelo, 12 October 1931, *CW*, 5:420–31. For examples of later attacks on the proof, see for example Perelman, "Paradoxes de la Logique" and the reply by Helmer, "Perelman *versus* Gödel."

44. Van Atten, "Gödel and Brouwer," 168.

45. Emil L. Post to KG, 29 October 1938, *CW*, 5:169.

46. Postscript added by KG in 1964 to a reprint of his 1934 lectures on undecidable propositions, *CW*, 1:369–70.

47. Post, "Finite Combinatory Processes"; Turing, "On Computable Numbers."

48. Report by Hans Hahn, 1 December 1932, reproduced in *GA*, 119.

49. Sigmund, "Dozent Gödel," 79–81.

50. Sigmund, "Dozent Gödel," 81–82; Edmund Hlawka interview, *W&B*, 238; *Kollegiengeld* receipt, calendar year 1937, KGP, 13c/44.

51. Karl Menger to KG, [1932], *CW*, 5:94–95; KG to Menger, 4 August 1932, *CW*, 5:96–97; Einnahmen, 1./X.1932–20./VIII 1933, KGP, 13b/31.

52. Rudolf Gödel, "Chronik der Familie," 58.

53. Herbert Feigl to KG, 23 November 1932, KGP, 1c/45.

54. Beller, *History of Austria*, 222; Sigmund, "Dozent Gödel," 83; KG's membership card in the Vaterländische Front reproduced in *GA*, 48.

55. Karl Menger to Oswald Veblen, 27 October 1933, Veblen, Papers, 8/10.

56. Menger, *Reminiscences*, 211; invitation from H. Hahn, Veblen, Papers, 6/13; KG, "Zur intuitionistischen Arithmetik."

57. Oswald Veblen to Karl Menger, 11 November 1932, Veblen, Papers, 8/10.

58. KG to Oswald Veblen, 11 January (cable), 25 January, and 31 March 1933, and Veblen to KG, 20 April 1933, Veblen, Papers, 6/5.

59. Taussky-Todd, "Remembrances of Gödel," 32.

60. Herbert Feigl to KG, 9 December 1933, KGP, 1c/45.

CHAPTER 6: THE SCHOLAR'S PARADISE

1. Taussky-Todd, "Remembrances of Gödel," 32; Manifest of Alien Passengers for the United States Immigration Officer at Port of Arrival, SS *Beregaria*, Passengers sailing from Cherbourg, 23 September 1933.
2. Manifest of Alien Passengers for the United States Immigration Officer at Port of Arrival, SS *Aquitania*, Passengers sailing from Cherbourg, 30 September 1933, Arriving Port of New York, 6 October 1933; Abraham Flexner to Edgar Bamberger, 26 September 1933, IAS, Faculty Files, Pre-1953.
3. Dyson, *Turing's Cathedral*, 24; Stern, "History of Institute," 1:24–25.
4. Klári von Neumann quoted in Dyson, *Turing's Cathedral*, 24.
5. Stern, "History of Institute," 1:26, 47–48.
6. Stern, "History of Institute," 1:48–49.
7. Stern, "History of Institute," 1:56.
8. Stern, "History of Institute," 1:2–3, 8.
9. Dyson, *Turing's Cathedral*, 28; Stern, "History of Institute," 1:15–17.
10. Stern, "History of Institute," 1:1–3.
11. Stern, "History of Institute," 1:4.
12. Flexner, "Useless Knowledge"; Stern, "History of Institute," 1:77–82, 134.
13. Oswald Veblen quoted in Dyson, *Turing's Cathedral*, 31–32.
14. Dyson, *Turing's Cathedral*, 32.
15. Dyson, *Turing's Cathedral*, 18–19, 21; Goldstine, OH, 4.
16. Oswald Veblen quoted in Dyson, *Turing's Cathedral*, 26; "A Memorial to a Scholar-Teacher," *Princeton Alumni Weekly*, 30 October 1931; Tucker, OH, 16.
17. Tucker, OH, 7–8.
18. Stern, "History of Institute," 1:73 n. 7, 139, 189 n. 19, 193 nn. 63 and 65, 194 n. 85.
19. Stern, "History of Institute," 1:195–96 n. 111.
20. Stern, "History of Institute," 1:219.
21. KG to MG, 29 September 1950; he mentions having written to her back in 1933 about "the local beauties of nature" in KG to MG, 28 May 1961.
22. Fitzgerald, *This Side of Paradise*, 40, 47.
23. Maynard, "Princeton in Confederacy's Service"; Watterson, *I Hear My People*, 79.

24. Albert Einstein to Elisabeth of Belgium, 20 November 1933, Einstein, Archives, 32-369.

25. Carl Ludwig Siegel to Richard Courant, 18 September 1935, quoted in Siegmund-Schultze, *Mathematicians Fleeing Nazi Germany*, 247.

26. Kreisel, "Kurt Gödel," 154.

27. Dawson, *Logical Dilemmas*, 97–98; KG to MG, 3 October 1948.

28. Albert Einstein to Hans Reichenbach, 2 May 1936, quoted in Siegmund-Schultze, *Mathematicians Fleeing Nazi Germany*, 226 (I have slightly adjusted the translation from the original German); Blaschke, "History of American Mathematical Society"; Segal, "Mathematics and German Politics," 131–32.

29. Albert Einstein quoted in Siegmund-Schultze, *Mathematicians Fleeing Nazi Germany*, 225; Birkhoff, "American Mathematics," 2:277; Abraham Flexner to George Birkhoff, 30 September 1938, quoted in Institute for Advanced Study, *Refuge for Scholars*, 8, and Siegmund-Schultze, *Mathematicians Fleeing Nazi Germany*, 226.

30. Halperin, OH, 14; Klári von Neumann quoted in Dyson, *Turing's Cathedral*, 54; Graham, "Adventures in Fine Hall."

31. Whitman, *Martian's Daughter*, 16–17.

32. Dyson, *Turing's Cathedral*, 54; Churchill Eisenhart quoted in Graham, "Adventures in Fine Hall."

33. Graham, "Adventures in Fine Hall."

34. KG to Oswald Veblen, 31 March 1933, Veblen, Papers, 6/5.

35. KG, "Foundation of Mathematics," *CW*, 3:45.

36. KG, "On Undecidable Propositions," CW, 1:346–71 (lectures at IAS); KG, "Existence of Undecidable Propositions" (NYU talk); Engelen, ed., *Notizbücher*, 394 (depressed afterward); John Kemeny quoted in Graham, "Adventures in Fine Hall." No notes of his Washington talk survive.

37. Siegmund-Schultze, *Mathematicians Fleeing Nazi Germany*, 244.

38. Abraham Flexner to KG, 7 March 1934, IAS, Faculty Files, Pre-1953.

39. *GA*, 51.

40. Sigmund, *Exact Thinking*, 291–92, 296; Menger, *Reminiscences*, 214.

41. Menger, *Reminiscences*, 211, 213.

42. Menger, *Reminiscences*, 211–12; Feferman and Feferman, *Tarski*, 81–82.

43. Feferman and Feferman, *Tarski*, 12, 19, 37–39, 53.

44. Feferman and Feferman, *Tarski*, 5, 84–85, 144–45.

45. Karl Menger quoted in Sigmund, *Exact Thinking*, 230; Morgenstern, "Abraham Wald," 361–62; OMD, 6 June 1936.

46. Menger, *Reminiscences*, 212–13; Morgenstern, "Abraham Wald," 363.

47. Menger, *Reminiscences*, 214.

48. Menger, *Reminiscences*, 216.

49. Kreisel, "Kurt Gödel," 154.

50. KG to Oswald Veblen, 1 January 1935, Veblen, Papers, 6/5.

51. Topp, "Hoffmann's Purkersdorf."

52. KG to MG, 28 July 1946.

53. Kreisel, "Kurt Gödel," 154; Dr. Philip Erlich File on Kurt Gödel, 14 April 1970, KGP, 27/1; Menger, *Reminiscences*, 211.

54. KG to MG, 19 September 1946, 11 January 1948, 10 March 1952.

55. Receipt, Dr. Max Schur, KGP, 13b/45; prescription, 4 August 1930, Dr. Otto Porges, KGP, 14a/14; KG to MG, 12 June 1952, 4 July 1962.

56. KG, "Protokolle," 11–12; Engelen, ed., *Notizbücher*, 283–87, 327–38, 400, 412.

57. Engelen, ed., *Notizbücher*, 504, 511, 513, 526.

58. Goldstein, *Incompleteness*, 215.

59. Hao Wang and Hassler Whitney interviews, *W&B*, 242, 250.

60. Huber-Dyson, "Gödel and Mathematical Truth."

61. KG to Oswald Veblen, 1 August and 17 November 1935, IAS, Faculty Files, Pre-1953.

62. Dawson, *Logical Dilemmas*, 109.

63. Abraham Flexner to KG, 18 and 21 November 1935, IAS, Faculty Files, Pre-1953; hotel receipts, Hotel Pennsylvania, Shelton Hotel, and Hotel New York, November 1935, KGP, 13b/24.

64. Although Rudolf Gödel years later gave conflicting accounts of whether he actually did go to Paris, Gödel's hotel bill shows that he arrived on December 7 and departed December 11, with a second room charged for December 9 to 11, suggesting that Rudi did indeed join him there to accompany him home. On December 8 Gödel ran up a telephone charge of 701 francs, the next day 102 francs—the total equaling about $50, thirty times the daily room charge: receipt, Palace Hotel, Paris, 7–11 December 1935, KGP, 13b/24.

65. Oswald Veblen to KG, 3 December 1935, Veblen, Papers, 6/5.

66. Oswald Veblen to Paul Heegaard, 10 December 1935, Veblen, Papers, 6/5.

67. Karl Menger to Oswald Veblen, 17 December 1935, Veblen, Papers, 8/10.

68. Moritz Schlick to [Otto Pötzl], 8 January 193[6], Schlick, Nachlass, Korrespondenz, 124/N.N-28. I am grateful to Karl Sigmund for providing me with a copy of this letter and for the identification of Pötzl as the unnamed "Nervenarzt" it was addressed to. Although the letter is dated

8 January 1935, it refers to events that definitively point to that being a mistake for 1936.

69. Frank Aydelotte to Selective Service Board, 19 May 1943, IAS, Visa-Immigration; Dr. Philip Erlich File on Kurt Gödel, 21 April 1970, KGP, 27/1.

70. Dr. Philip Erlich File on Kurt Gödel, 21 May 1970, KGP, 27/1.

CHAPTER 7: FLEEING THE REICH

1. OMD, 4 July 1940 ("talkative"), 18 November 1944 ("schrecklich"), 12 July 1940 ("fürchterlich"), 12 February 1947 ("grässlich"), 7 March 1948 ("abscheulich"), 11 December 1947 ("nicht erfreulich"), 16 July 1949 ("solche Pest").

2. Karl Menger to Oswald Veblen, n.d. 1938, Veblen, Papers, 8/10.

3. Frank Aydelotte to Selective Service Board, 19 May 1943, IAS, Visa-Immigration.

4. OMD, 12 July 1940, 11 December 1947.

5. Engelen, ed., *Notizbücher*, 385, 452, 487; KG, "Protokolle," 10, 17. In the phrase "Sadismus und Sonstiges (keine reine Liebe)," *reine* seems the likeliest reading for the barely legible shorthand word between *keine* and *Liebe*.

6. OMD, 3 March 1947, 7 October 1941.

7. OMD, 7 March 1946, 2 October 1945, 12 July 1940.

8. Kreisel, "Kurt Gödel," 153, 154–56; Kreisel, "Gödel's Intuitionist Logic," 150.

9. Huber-Dyson, "Gödel and Mathematical Truth."

10. Alfred Tarski to Adele Gödel, 4 January 1943, quoted in Feferman and Feferman, *Tarski*, 145, 152; Tarski to KG and Adele Gödel, 9 December 1943, ibid., 152.

11. OMD, 2 July 1945; in OMD, 10 February 1970, Morgenstern added that von Neumann had visited Gödel in the hospital or sanatorium specifically so that Gödel could explain to him his proof.

12. Jones, "Insulin Coma Therapy"; Dr. Philip Erlich File on Kurt Gödel, 21 April 1970, KGP, 27/1.

13. The account of Schlick's murder here and below is drawn from the official documents of the case translated and reprinted in Stadler, *Vienna Circle*, 600–630, with additional details in Sigmund, *Exact Thinking*, 283–86, 312–17.

14. Menger, *Reminiscences*, 196–97.

15. "The Case of Professor Schlick in Vienna—A Reminder to Search our

Conscience, by Prof. Dr. Austriacus," English translations in Stadler, *Vienna Circle*, 602–6, and Stadler and Weibel, eds., *Vertreibung der Vernuft*, 15. For background on *Die schönere Zukunft* and *Das neue Reich*, see Wasserman, *Black Vienna*, 193–94.

16. Gödel's typescript copy is in KGP, 14c/11.

17. Clemency Petition to Ministry of Justice from Dr. Joh. Sauter, reprinted in Stadler, *Vienna Circle*, 624–25.

18. Nagel, "Philosophy in Europe II," 30 n. 2.

19. Receipts, Purkersdorf Sanatorium, May–September 1936, KGP, 13c/45; hotel receipt, Aflenz, 3 October 1936, KGP, 13b/25; OMD, 13 March 1970.

20. Menger, *Reminiscences*, 215; Karl Menger interview, *W&B*, 230.

21. Karl Menger to Franz Alt, 31 December 1937, quoted in Sigmund, *Exact Thinking*, 320.

22. Menger, *Reminiscences*, 215–16.

23. Karl Popper quoted in Sigmund, *Exact Thinking*, 322; Taussky-Todd, OH, 30.

24. Notes of conversation with Olga Taussky, 21 September 1937, KG, "Protokolle," 20–21, 24.

25. KG, "Protokolle," 31.

26. Rudolf Gödel, "Chronik der Familie," 59.

27. KG, "Protokolle," 17–18.

28. KG to Karl Menger, 3 July 1937, *CW*, 5:106–8.

29. Notes of conversation with John von Neumann, 17 July 1937, KG, "Protokolle," 6–7; Walker, "German Physics," 65; Ball, *Serving the Reich*, 91–92, 97–98.

30. Oswald Veblen to KG, 1 November 1937, IAS, Faculty Files, Pre-1953.

31. KG to Karl Menger, 15 December 1937, *CW*, 5:112–14.

32. Engelen, ed., *Notizbücher*, 354, 414, 423, 439, 479.

33. KG, "Protokolle," 18.

34. Simotta, "Marriage and Divorce in Austria," 525.

35. Invoices from Georg Rathauscher, Elektriker, 17 November 1937, and W. Nekola, Spenglerei, 17 November 1937, KGP, 13b/14.

36. KG, "Lecture at Zilsel's"; KG, "Protokolle," 31–33, 57–60, 75.

37. KG, "Protokolle," 7, 32–34.

38. Abraham Flexner to KG, 21 February 1938, IAS, Faculty Files, Pre-1953.

39. Berkley, *Vienna and Its Jews*, 252.

40. Zuckmayer, *Part of Myself*, 49; Beller, *History of Austria*, 229–30.

41. Zuckmayer, *Part of Myself*, 50.

42. Zuckmayer, *Part of Myself*, 52; Berkley, *Vienna and Its Jews*, 259–60.

43. Alfred Polgar quoted in Clare, *Last Waltz in Vienna*, 221; de Waal, *Hare with Amber Eyes*, 241–42.

44. Rudin, *Way I Remember*, 33.

45. Rudin, *Way I Remember*, 36; Sigmund Freud quoted in Stadler and Weibel, eds., *Vertreibung der Vernuft*, 366.

46. Stadler and Weibel, eds., *Vertreibung der Vernuft*, 64; Sigmund, *Exact Thinking*, 325; Viktor Christian to KG, 23 April 1938, KGP, 13a/3.

47. "Viktor Christian, Prof. Dr.," University of Vienna, *650 Plus*.

48. National Archives, Applications to NS-Frauenschaft, Hildegarde Porkert, roll E010; Nazi Party Applications by Austrians, Josef Porkert, roll 846.

49. Dr. Philip Erlich File on Kurt Gödel, 1, 8, and 15 March 1977, KGP, 27/1.

50. Simotta, "Marriage and Divorce in Austria," 526 and n. 6.

51. Power of Attorney, 29 August 1938, KGP, 13a/13; receipt, Rathauskeller, 20 September 1938, KGP, 13b/21.

52. Karl Menger to Oswald Veblen, [n.d., 1938], Veblen, Papers, 8/10; Karl Menger to KG, [December 1938], *CW*, 5:125.

53. KG to Karl Menger, 25 June, 19 October, and 11 November 1938, quoted in Menger, *Reminiscences*, 218–19.

54. Menger, *Reminiscences*, 220–21.

55. Menger, *Reminiscences*, 224.

56. Menger, *Reminiscences*, 224–25; KG to Karl Menger, 30 August 1939, CW, 5:124–26; OMD, 19 March 1972.

57. KG to Oswald Veblen, draft letter, November 1939, KGP, 13c/197.

58. John von Neumann to Abraham Flexner, 27 September 1939, quoted in Dyson, *Turing's Cathedral*, 96; von Neumann to KG, telegram, 5 October 1939, IAS, Faculty Files, Pre-1953.

59. John von Neumann to Abraham Flexner, 16 October 1939, IAS, Visa-Immigration.

60. Ash, "Universität Wien," 124–25; Friedrich Plattner to Rektor der Universität, 12 August 1939, reproduced in *GA*, 67–68.

61. Arthur Marchet to Rektor der Universität, 30 September 1939, reproduced in *GA*, 72.

62. Dawson, *Logical Dilemmas*, 140; KG to Devisenstelle Wien, 29 July 1939, reproduced in *GA*, 65–66.

63. KG to Oswald Veblen, draft letter, November 1939, KGP, 3c/197; Menger, *Reminiscences*, 224; Kreisel, "Kurt Gödel," 155.

64. Frank Aydelotte to Chargé d'Affaires, German Embassy, 1 December 1939, IAS, Faculty Files, Pre-1953.

65. Der Dekan to Rektor der Universität, 27 November 1939, reproduced in *GA*, 71.

66. KG to Frank Aydelotte, 5 January 1940, IAS, Faculty Files, Pre-1953; KG to Institute for Advanced Study, telegram, 15 January 1940, ibid.; KG passport, KGP, 13a/8.

67. KG to MG, 29 November 1965 ("I still recall the suitcase of things that Adele brought back from there in 1940").

68. KG to RG, 31 March 1940; KG to Institute for Advanced Study, telegram, 5 March 1940, IAS, Faculty Files, Pre-1953.

69. OMD, 10 March 1940.

CHAPTER 8: NEW WORLDS

1. Kreisel, "Kurt Gödel," 157.

2. KG to RG, 31 March 1940.

3. "Allgemeine Bildung" notebooks, KGP, 5b/1–11.

4. KG to RG, 15 September and 5 June 1940.

5. Sigmund, *Exact Thinking*, 348–51.

6. KG to MG, 17 May 1946, 19 January 1947.

7. Abraham Flexner to Oswald Veblen, 6 January 1937, quoted in Dyson, *Turing's Cathedral*, 88; Flexner to Herbert Maass, 15 December 1937, quoted in ibid., 19, 88.

8. KG to MG, 4 May 1941 ("I only had 3 listeners left by the end").

9. KG, "Cantor's Continuum Problem," *CW*, 2:186.

10. Hao Wang interview, *W&B*, 241; Wang, *Reflections on Gödel*, 116, 131–32.

11. Kreisel, "Kurt Gödel," 158.

12. Schewe, *Freeman Dyson*, 119–22; Huber-Dyson, "Gödel and Mathematical Truth"; Kreisel, "Gödel's Excursions," 146; Feferman and Feferman, *Tarski*, 228.

13. Huber-Dyson, "Gödel and Mathematical Truth"; Schewe, *Freeman Dyson*, 122; Feferman and Feferman, *Tarski*, 273–76.

14. Graham, "Adventures in Fine Hall."

15. Holton and Elkana, eds., *Albert Einstein*, 4.

16. Pais, *Subtle Is the Lord*, 473; OMD, 7 December 1947.

17. KG to MG, 21 July 1946; Dyson, *Gaia to Eros*, 161; Ernst Straus in Holton and Elkana, eds., *Albert Einstein*, 422; Straus in Woolf, ed., *Strangeness in Proportion*, 485.

18. KG to Carl Seelig, 7 September 1955, *CW*, 5:249; OMD, 7 December 1947, 10 February 1951.

19. KG to MG, 31 July 1947, 16 April 1949, 26 February 1949, 5 January 1955.

20. KG to MG, 31 July 1947.

21. OMD, 11 June 1946; KG to MG, 19 January 1947.

22. OMD, 29 March 1946, 20 September 1958, 18 July 1969.

23. OMD, 9 December 1969.

24. OMD, 7 October 1941.

25. OMG to MG, 22 January 1946; OMG to RG, 21 September 1941; Frank Aydelotte to Dr. Max Gruenthal, 5 December 1941, IAS, Faculty Files, Pre-1953.

26. Louise Morse interview, *W&B*, 251.

27. KG to RG, 4 May 1941; Dawson, *Logical Dilemmas*, 160.

28. KG to RG, 16 March 1941.

29. Diploma "Im Namen des führers," reproduced in *GA*, 83.

30. KG to RG, 6 October 1940.

31. Reich Ministry for Science, Education, and Culture to Herr Rektor der Universität in Wien, 17 July 1941, reproduced in *GA*, 84.

32. Frank Aydelotte to Dr. Max Gruenthal, 2 and 5 December 1941, and Gruenthal to Aydelotte, 4 December 1941, IAS, Faculty Files, Pre-1953.

33. KG to Commissioner, Immigration and Naturalization Service, 1 December 1943, IAS, Visa-Immigration.

34. Frank Aydelotte to Selective Service Board, 14 April and 19 May 1943, IAS, Visa-Immigration.

35. Ceville O. Jones, Selective Service Board, to Frank Aydelotte, 20 April 1943, IAS, Visa-Immigration; KG to MG, 4 July 1962.

36. KG, "New Cosmological Solutions"; KG, "Rotating Universes." Gödel's theory is explained in Bonnor, *Expanding Universe*, 144–54.

37. KG, "Relativity and Idealistic Philosophy," *CW*, 2:202.

38. KG, "Relativity and Idealistic Philosophy," *CW*, 2:205 and n. 11.

39. KG, "Relativity and Idealistic Philosophy," *CW*, 2:205–6.

40. Kreisel, "Kurt Gödel," 155.

41. KG to MG, 7 September 1945.

42. KG to MG, 7 November 1947.

43. Bryant, *Prague in Black*, 227; Zimmermann, *Sudetendeutsch*, 135; Rudolf Gödel, "Chronik der Familie," 59–60.

44. Bryant, *Prague in Black*, 237–38.

45. KG to MG, 7 February 1957, 11 September 1960.

46. RG to KG, 21 January 1946, IAS, Faculty Files, Pre-1953.

47. OMD, 7 March 1946; KG to MG, 19 September 1946, 12 May 1947; Adele Gödel to RG, 3 January 1948.

48. KG to MG, 12 and 26 May 1947.
49. Rudolf Gödel, "History of Gödel Family," 27.
50. KG to MG, 6 July 1960.
51. KG to MG, 12 and 26 May 1947.
52. OMD, 11 July 1947.
53. Notes on American Government and History, KGP, 11b/1–2; OMD, 23 February 1947.
54. Oskar Morgenstern, "History of the Naturalization of Kurt Gödel," draft memorandum, 13 September 1971, in Thomas, Collection.
55. Guerra-Pujol, "Gödel's Loophole."
56. John von Neumann to Oswald Veblen, 30 November 1945, quoted in Dyson, *Turing's Cathedral*, 100.
57. KG to MG, 28 April 1946.

CHAPTER 9: PLATO'S SHADOW

1. Joseph M. Rampona interview, *W&B*, 247.
2. OMD, 10 February 1951; KG to J. Robert Oppenheimer, n.d. [February 1951], IAS, Faculty Files, Pre-1953.
3. Oppenheimer to KG, 15 February 1951, IAS, Faculty Files, Pre-1953.
4. KG to RG, 18 March 1951.
5. OMD, 10 February 1951.
6. OMD, 24 February 1951; Dawson, *Logical Dilemmas*, 194.
7. OMD, 12 and 14 March 1951; KG to MG, 12 April 1951.
8. "World News Summarized," *New York Times*, 12 March 1951; KG to MG, 13 May and 28 June 1951.
9. Russell, *Autobiography, 1914–1944*, 341.
10. KG to Kenneth Blackwell, unsent reply to letter of 22 September 1971, *CW*, 4:316–17; Russell, *Mathematical Philosophy*, 169; KG, "Russell's Logic," *CW*, 2:120.
11. KG, "Basic Theorems and Implications," *CW*, 3:314.
12. KG, "Basic Theorems and Implications," *CW*, 3:310.
13. KG, "Basic Theorems and Implications," *CW*, 3:311–12.
14. KG, "Basic Theorems and Implications," *CW*, 3:321–22.
15. KG, "Basic Theorems and Implications," *CW*, 3:323.
16. OMD, 12 May 1963, 20 November 1971; Rudolf Carnap, notes of conversation with KG, 13 November 1940, quoted in Gierer, "Gödel Meets Carnap," 213; Wang, *Logical Journey*, 167.
17. Wang, *Logical Journey*, 8.

18. Crocco and Engelen, "Gödel's Philosophical Remarks," 36. I have slightly adjusted their translation of no. 14; the original German text reads, "Die Religionen sind zum größten Teil schlecht, aber nicht die Religion."

19. Paul Erdös quoted in Regis, *Einstein's Office*, 64.

20. KG to MG, 17 December 1948 ("forgeries"); 21 September 1953 ("1:2000"); 12 December 1956 (world wars); 27 September 1951 ("sabotage"); 27 July 1951 ("not learn that"); 5 January 1947 ("no further").

21. KG to MG, 22 July 1952.

22. Russell, *Western Philosophy*, 581.

23. OMD, 30 October 1945, 17 September 1944.

24. Menger, *Reminiscences*, 222–23.

25. OMD, 27 June 1945.

26. OMD, 11 June 1946.

27. OMD, 5 October 1953, 27 June 1945.

28. Menger, *Reminiscences*, 226–27.

29. KG to RG, 5 January 1955; KG to MG, 7 November 1956.

30. KG to MG, 9 June 1948, 5 January 1947.

31. KG to MG, 30 July 1950.

32. KG to MG, 8 January 1951; Cornelius A. Moynihan to J. Edgar Hoover, 23 September 1952, reproduced in *GA*, 145.

33. KG to MG, 16 January 1956, 22 November 1946, 17 March 1951.

34. KG to MG, 22 July 1952.

35. Ernest Straus in Woolf, ed., *Strangeness in Proportion*, 485.

36. OMD, 6 November 1960, 27 February 1972; "Colleagues Back Dr. Oppenheimer," *New York Times*, 1 July 1954; KG to MG, 12 December 1956; Paul Erdös interview, *W&B*, 235.

37. Ulam, *Adventures of a Mathematician*, 80; Carl Siegel quoted in Dawson, *Logical Dilemmas*, 194.

38. KG to MG, 31 October 1953; Atle Selberg quoted in Regis, *Einstein's Office*, 64.

39. Regis, *Einstein's Office*, 206; Selberg, OH, 32–33.

40. Selberg, Papers, 11/5; J. Robert Oppenheimer to Atle Selberg, 2 and 3 November 1964, Weil, Faculty Files.

41. "Dispute Splits Advanced Study Institute," *New York Times*, 2 March 1973; Jones, "Mount Olympus"; Goldstein, *Incompleteness*, 245; KG to MG, 30 September 1956.

42. KG to MG, 16 June 1946, 16 March 1947.

43. OMD, 12 July 1940; KG to MG, 16 June 1946 (appendix operation), 15 August 1946 (teeth), 9 June 1948 (weight).

44. Freeman Dyson, diary entry 25 November 1948 quoted in Dyson, *Maker of Patterns*, 126–27.

45. Dr. Philip Erlich File on Kurt Gödel, 21 May 1970, KGP, 27/1.

46. KG to MG, 14 April and 10 May 1953.

47. OMD, 16 July 1949; Mrs. E. W. Leary to J. Robert Oppenheimer, 20 September 1950, IAS, Faculty Files, Pre-1953.

48. KG to MG, 11 September 1949; KG to RG, 30 July 1950.

49. Kreisel, "Gödel's Excursions," 146; KG to MG, 16 April 1949, 8 January 1951.

50. KG to MG, 7 June 1959 (flamingo), 20 September 1952 and 23 August 1953 (Skee-Ball).

51. KG to MG, 16 January 1956 (*Aïda*); Rudolf Gödel, "History of Gödel Family," 26 (modern painting); KG to MG, 30 April 1961 (*Hamlet*), 28 May 1961 (Gogol).

52. KG to MG, 4 July 1962 (Kafka), 30 December 1950 (Wagner), 16 December 1960 (Mahler); KG to MG, 26 July 1953 ("Bach and Wagner").

53. Wang, *Reflections on Gödel*, 218–19; KG to MG, 25 April 1955; Selberg, Papers, 11/5.

54. Abramowitz, Schwartz, and Whiteside, "Conceptual Model of Hypochondriasis."

55. Fiebertabellen, KGP, 14a/15; Kreisel, "Kurt Gödel," 152–53.

56. KG prescriptions, KGP, 4a/14; *GA*, 99.

57. "Bland Diet Instructions," KGP, 14a/16; KG to MG, 5 January 1955, 30 June 1951, 12 January 1958, 15 January 1959.

58. OMD, 7 February, 25 November, and 4 December 1954, 9 January 1955, 9 November 1957, 6 November 1960, 22 December 1963, 18 May 1968, 10 February, 12 March, and 29 August 1970.

59. Albert Einstein to Elsa Löwenthal, ca. 11 August 1913 (no. 466), Einstein, *Collected Papers*, 5 (English trans. supp.): 348; Pais, *Subtle Is the Lord*, 477.

60. KG to MG, 25 April and 21 June 1955.

61. Sacks, "Reflections on Gödel"; KG to John von Neumann, 20 March 1956, CW, 5:373–77; Wigderson, "Gödel Phenomenon."

62. CW, 5:335.

63. KG to MG, 9 June 1948, 12 November 1951.

64. KG to MG, 9 August 1957.

65. "Viktor Christian, Prof. Dr.," University of Vienna, *650 Plus*; "Fritz (Friedrich) Knoll, Prof. Dr.," ibid.; Pfefferle and Pfefferle, *Glimpflich entnazifiert*, 154–55; "The De-Nazification of the Professorate at the University of Vienna, 1945–1950," University of Vienna, *650 Plus*.

66. Köhler, "Philosophy of Misdeed"; Sigmund, *Exact Thinking*, 360–61.

67. Köhler, "Philosophy of Misdeed."

68. Sigmund, *Exact Thinking*, 356–57.

69. KG to MG, 11 December 1946.

70. KG to MG, 29 March 1956.

71. KG to MG, 8 May and 7 June 1958; KG to RG, 7 June 1958.

CHAPTER 10: "IF THE WORLD IS CONSTRUCTED RATIONALLY"

1. "Interview with Martin Davis," 567.

2. Paul J. Cohen to KG, 24 April 1963, KGP, 1b/27.

3. Paul J. Cohen to KG, 6 May 1963, KGP, 1b/27.

4. KG to Paul J. Cohen, 5 June 1963, CW, 4:378 and 20 June 1963, CW, 4:382–83.

5. KG to Alonzo Church, 10 August 1966, CW, 4:371–72.

6. Sacks, "Reflections on Gödel."

7. KG, "Modern Development of Foundations," CW, 3:375, 381.

8. Wang, *Logical Journey*, 192–93; Goldstein, *Incompleteness*, 31–32; KG, "Modern Development of Foundations," CW, 3:375.

9. KG, "Modern Development of Foundations," 379, 381.

10. KG, "Cantor's Continuum Problem," supplement to second edition, CW, 2:268.

11. KG, "Cantor's Continuum Problem," supplement to second edition, CW, 2:268–69; KG, "Modern Development of Foundations," CW, 3:385.

12. KG, "Modern Development of Foundations," CW, 3:381, 383.

13. KG to MG, 23 July and 6 October 1961.

14. KG to MG, 6 October and 12 September 1961.

15. OMD, 4 August and 29 August 1970; CW, 3:405, 424.

16. OMD, 29 August 1970.

17. CW, 5:135–44.

18. IAS, School of Mathematics, Faculty Files.

19. Interview requests, KGP, 4c/50; Elinn Definbaugh to KG, [1] December 1971, KGP, 4c/2; crank correspondence, KGP, 14a/18–20; Sacks, "Reflections on Gödel."

20. "'Even Princeton' Travels to Protest: SDS Attends Washington Rally," *Daily Princetonian*, 30 November 1965, 1; KG to MG, 21 October 1965.

21. "Brick is Hurled," *Daily Princetonian*, 7 May 1970, 5; "Protesters Yield to Restraining Order," *Daily Princetonian*, 12 May 1970, 1.

22. "Valentines," *Princeton Town Topics*, 10 February 1972, 33.

23. OMD, 15 December 1971, 1 January 1972, 11 October 1973.
24. OMD, 7 April 1974.
25. OMD, 17 and 20 September 1975.
26. OMD, 23 February 1976, 6 September 1975.
27. OMD, 6 February 1970, 12 April 1974.
28. OMD, 9 April 1974.
29. OMD, 11 July 1977; Oskar Morgenstern memorandum on Kurt Gödel, 10 July 1977, quoted in Dawson, *Logical Dilemmas*, 251; Thomas, OH, 13.
30. Dr. Philip Erlich File on Kurt Gödel, KGP, 27/1; OMD, 12 February 1976, 1 January 1977, 11 June 1976.
31. Dr. Philip Erlich File on Kurt Gödel, 12 June 1970, 28 August 1977, KGP, 27/1.
32. Adeline Federici interview, *W&B*, 246; Dawson, *Logical Dilemmas*, 252; Wang, *Reflections on Gödel*, 133.
33. Minot C. Morgan, Jr., to Dr. Woolf et al., 19 December 1977, IAS, Faculty Files, 1951–1977; Wang, *Reflections on Gödel*, 133.
34. Wang, *Reflections on Gödel*, 133.
35. Samuel M. Kind to Adele Gödel, 3 March 1978, and Homer R. Zink to Carl Pope, 14 March 1978, IAS, Faculty Files, 1978–1981.
36. Homer R. Zink to Adele Gödel, 2 March 1978, ibid.; Dawson, *Logical Dilemmas*, 258.
37. John Dawson, OH, 3–8.
38. Sokal and Bricmont, *Fashionable Nonsense*, 176; Ellenberg, "Does Gödel Matter?"; Raatikainen, "Relevance of Gödel's Theorems," 528–29.
39. KG to MG, 20 October 1963.
40. Memorandum on KG's "Work," reproduced in *GA*, 151.
41. Wigderson, "Gödel Phenomenon," 475–76; Hoffmann, *Grenzen der Mathematik*, 52–53.
42. Fogelin, *Tightrope of Reason*, 97.
43. Sacks, "Reflections on Gödel."

BIBLIOGRAPHY

WORKS BY KURT GÖDEL

Kurt Gödel Papers. Institute for Advanced Study Archives. On deposit at the Manuscripts Division, Special Collections, Princeton University Library, Princeton, N.J.

Letters to Marianne and Rudolf Gödel. Wienbibliothek im Rathaus, Vienna. Digital publication, digital.wienbibliothek.at.

"Protokolle." Gabelsberger shorthand records of conversations and personal reflections, 1937–1938. Kurt Gödel Papers, box 6c, folder 81. Digital publication of Gabelsberger transcription by Erich Ruff with English translation by Marilya Veteto Reese, budiansky.com/goedel.

Collected Works. Edited by Solomon Feferman et al. 5 vols. New York: Oxford University Press, 1986–2003.

"Über die Vollständigkeit des Logikkalküls" [On the Completeness of the Calculus of Logic]. PhD dissertation, University of Vienna, 1929. *Collected Works*, 1:60–101.

Lecture at Königsberg on Completeness of the Functional Calculus. Presented at the Second Conference on Epistemology of the Exact Sciences, Königsberg, September 5–7, 1930. *Collected Works*, 3:16–29.

"Die Vollständigkeit der Axiome des logischen Funcktionenkalküls" [The Completeness of the Axioms of the Functional Calculus of Logic]. *Monatshefte für Mathematik und Physik* 37 (1930): 349–60. *Collected Works*, 1:103–23.

"Über formal unentscheidbare Sätze der *Principia mathematica* und verwandter Systeme I" [On Formally Undecidable Propositions of *Principia Mathe-*

matica and Related Systems]. *Monatshefte für Mathematik und Physik* 38 (1931): 173–98. *Collected Works*, 1:144–95.

"Zur intuitionistischen Arithmetik und Zahlentheorie" [On Intuitionistic Arithmetic and Number Theory]. *Ergebnisse eines mathematischen Kolloquiums* 4 (1933): 34–38. Collected *Works*, 1:286–95.

"The Present Situation in the Foundations of Mathematics." Paper presented at meeting of the Mathematical Association of America, Cambridge, Mass., December 29–30, 1933. *Collected Works*, 3:45–53.

"On Undecidable Propositions of Formal Mathematical Systems." Notes of lectures at the Institute for Advanced Study, February–May 1934. *Collected Works*, 1:346–71.

"The Existence of Undecidable Propositions in any Formal Mathematical System Containing Arithmetic." Manuscript notes of lecture to Philosophical Society of New York University, 18 April 1934. Kurt Gödel Papers, box 7b, folder 30. Digital publication, budiansky.com/goedel.

"Lecture at Zilsel's." *Collected Works*, 3:86–113.

"The Consistency of the Axiom of Choice and of the Generalized Continuum Hypothesis." *Proceedings of the National Academy of Sciences* 24 (1938): 556–57. *Collected Works*, 2:26–27.

"Consistency Proof for the Generalized Continuum Hypothesis." *Proceedings of the National Academy of Sciences* 25 (1939): 220–24. *Collected Works*, 2:28–32.

"The Consistency of the Axiom of Choice and of the Generalized Continuum Hypothesis with the Axioms of Set Theory." *Annals of Mathematical Studies*, vol. 3. Princeton: Princeton University Press, 1940. *Collected Works*, 2:33–101.

"In What Sense Is Intuitionistic Logic Constructive?" Lecture at Yale University, April 15, 1941. *Collected Works*, 3:189–200.

"Russell's Mathematical Logic." In *The Philosophy of Bertrand Russell*, edited by Paul A. Schlipp. Library of Living Philosophers, vol. 5. Evanston, Ill.: Northwestern University Press, 1944. *Collected Works*, 2:119–43.

"What Is Cantor's Continuum Problem?" *American Mathematical Monthly* 54 (1947), 515–25. *Collected Works*, 2:176–88.

"An Example of a New Type of Cosmological Solutions of Einstein's Field Equations of Gravitation." *Reviews of Modern Physics* 21 (1949): 447–50. *Collected Works*, 2:190–98.

"A Remark about the Relationship between Relativity Theory and Idealistic Philosophy." In *Albert Einstein, Philosopher–Scientist*, edited by Paul A. Schlipp. Library of Living Philosophers, vol. 7. Evanston, Ill.: Northwestern University Press, 1949. *Collected Works*, 2:202–7.

"Rotating Universes in General Relativity Theory." *Proceedings of the International Congress of Mathematicians, Cambridge, Massachusetts, August 30–September 6, 1950*, 1:175–81. *Collected Works*, 2:208–16.

"Some Basic Theorems on the Foundations of Mathematics and their Implications." 25th Josiah Willard Gibbs Lecture, American Mathematical Society, Providence, R.I., December 26, 1951. *Collected Works,* 3:304–23.

"Is Mathematics Syntax of Language?" Unpublished essay, 1953–59. *Collected Works*, 3:334–62.

"The Modern Development of the Foundations of Mathematics in the Light of Philosophy." Draft of undelivered lecture, 1961. *Collected Works*, 3:374–87.

Unpublished ontological proof, 1970. *Collected Works*, 3:403–4.

OTHER SOURCES

Abramowitz, Jonathan S., Stefanie A. Schwartz, and Stephen P. Whiteside. "A Contemporary Model of Hypochondriasis." *Mayo Clinic Proceedings* 77 (2002): 1323–30.

Adani, Christoph. "Three Weeks with Hans A. Bethe." In *Hans Bethe and His Physics*, edited by Gerald Edward Brown and Chang-Hwan Lee. Singapore: World Scientific, 2006.

Aly, Götz. *Why the Germans? Why the Jews?* 2011. Translated by Jefferson Chase. New York: Picador, 2015.

Ash, Mitchell G. "Die Universität Wien in den politischen Umbrüchen des 19. und 20. Jahrhunderts." In *Universität–Politik–Gesellschaft*, edited by Mitchell G. Ash, Friedrich Stadler, and Josef Ehmer. Göttingen: V&R unipress, 2015.

Baaz, Matthias, et al., eds. *Kurt Gödel and the Foundations of Mathematics: Horizons of Truth.* Cambridge: Cambridge University Press, 2011.

Bahr, Hermann. *Wien.* Stuttgart: Carl Crabbe Verlag, 1906.

Ball, Philip. *Serving the Reich: The Struggle for the Soul of Physics under Hitler.* Chicago: University of Chicago Press, 2014.

Beller, Steven. *A Concise History of Austria.* Cambridge: Cambridge University Press, 2006.

———. *Vienna and the Jews, 1867–1938: A Cultural History.* Cambridge: Cambridge University Press, 1989.

Berend, Ivan. *Case Studies on Modern European Economy: Entrepreneurship, Inventions, and Institutions.* London: Routledge, 2013.

Berkley, George E. *Vienna and Its Jews: The Tragedy of Success, 1880s–1890s.* Lanham, Md.: Madison Books, 1988.

Birkhoff, George. "Fifty Years of American Mathematics." In *American Mathematical Society Semicentennial Publications*. 2 vols. New York: American Mathematical Society, 1938.

Blaschke, Wilhelm. Review of *Semicentennial History of the American Mathematical Society, 1888–1938*. *Jahresbericht der Deutschen Mathematiker-Vereinigung* 49 (1939): Pt. II, 80–81.

Bonnor, William. *The Mystery of the Expanding Universe*. New York: Macmillan, 1964.

Broch, Hermann. *The Unknown Quantity*. 1935. Translated by Willa and Edwin Muir. Marlboro, Vt.: Marlboro Press, 1988.

Brouwer, L. E. J. "Intuitionistic Reflections on Formalism." 1927. Reprinted in *From Frege to Gödel: A Source Book in Mathematical Logic, 1879–1932*, edited by Jean van Heijenoort. Cambridge: Harvard University Press, 1967.

Bryant, Chad. *Prague in Black: Nazi Rule and Czech Nationalism*. Cambridge: Harvard University Press, 2007.

Carnap, Rudolf. Papers. Archives and Special Collections, University of Pittsburgh Library, Pittsburgh, Pa.

Claims Resolution Tribunal of the Holocaust Victim Assets Litigation against Swiss Banks and other Swiss Entities. United States District Court for the Eastern District of New York. crt-ii.org.

Clare, George. *Last Waltz in Vienna*. London: Macmillan, 1981.

Connolly, P. J. "Karl Lueger: Mayor of Vienna." *Studies: An Irish Quarterly Review 5*, no. 14 (June 1915): 226–49.

Crocco, Gabriella, and Eva-Maria Engelen. "Kurt Gödel's Philosophical Remarks." In *Kurt Gödel: Philosopher–Scientist*, edited by Gabriella Crocco and Eva-Maria Engelen. Marseille: Presses Universitaires de Provence, 2016.

Curbera, Guillermo P. *Mathematicians of the World, Unite! The International Congress of Mathematicians—A Human Endeavor*. Boca Raton, Fla.: CRC Press, 2009.

Davis, Martin. "What Did Gödel Believe, and When Did He Believe It?" *Bulletin of Symbolic Logic* 11, no. 2 (June 2005): 194–206.

Dawson, John W., Jr. *Logical Dilemmas: The Life and Work of Kurt Gödel*. Wellesley, Mass.: A. K. Peters, 1997.

———. Oral History Project. Institute for Advanced Study Archives, Princeton, N.J.

———. "The Reception of Gödel's Incompleteness Theorem." *PSA: Proceedings of the Biennial Meeting of the Philosophy of Science Association, 1984*. Vol. 2: Symposia and invited papers (1984): 253–71.

Deak, John David. "The Austrian Civil Service in an Age of Crisis: Power and the Politics of Reform, 1848–1925." PhD dissertation, University of Chicago, 2009.

DePauli-Schimanovich, Werner. *Kurt Gödel und die Mathematische Logik.* Linz: Trauner Verlag, 2005.

DePauli-Schimanovich, Werner, and Peter Weibel. *Kurt Gödel – Ein mathematischer Mythos.* DVD video. Österreichischer Rundfunk, 1986.

De Waal, Edmund. *The Hare with the Amber Eyes.* New York: Picador, 2010.

Dyson, Freeman. *From Eros to Gaia.* New York: Penguin, 1995.

———. *Maker of Patterns: An Autobiography Through Letters.* New York: Liveright, 2018.

Dyson, George. *Turing's Cathedral: The Origins of the Digital Universe.* New York: Random House, 2012.

Einstein, Albert. The Albert Einstein Archives. Hebrew University of Jerusalem, Israel. albert-einstein.org.

———. *Collected Papers of Albert Einstein.* Digital Einstein Papers, Princeton University Press. einsteinpapers.press.princeton.edu.

———. Director's Office: Faculty Files. Institute for Advanced Study Archives, Princeton, N.J.

Ellenberg, Jordan. "Does Gödel Matter?" *Slate,* 10 March 2005.

Engelen, Eva-Maria, ed. *Kurt Gödel Philosophische Notizbücher/Philosophical Notebooks.* Vol. 2, *Zeiteinteilung (Maximen) I und II/Time Management (Maxims) I and II.* Berlin: De Gruyter, 2020.

Feferman, Anita Burdman, and Solomon Feferman. *Alfred Tarski: Life and Logic.* Cambridge: Cambridge University Press, 2004.

Feferman, Solomon. "Provenly Unprovable." Review of *Incompleteness* by Rebecca Goldstein. *London Review of Books,* February 9, 2006.

Feferman, Solomon, Charles Parson, and Stephen G. Simpson, eds. *Kurt Gödel: Essays for His Centennial.* Cambridge: Cambridge University Press, 2010.

Feigl, Herbert. "The Wiener Kreis in America." In *Perspectives in American History,* vol. 2: *The Intellectual Migration: Europe and America, 1930–1960,* edited by Donald Fleming and Bernard Bailyn. Cambridge, Mass.: Charles Warren Center for Studies in American History, 1968.

Fitzgerald, F. Scott. *This Side of Paradise.* New York: Scribner's, 1921.

Flexner, Abraham. "The Usefulness of Useless Knowledge." *Harper's Magazine,* October 1939, 544–52.

Fogelin, Robert J. *Walking the Tightrope of Reason: The Precarious Life of a Rational Animal.* New York: Oxford University Press, 2003.

———. "Wittgenstein and Intuitionism." *American Philosophical Quarterly* 5, no. 4 (October 1968): 267–74.

Fraenkel, Abraham A. *Recollections of a Jewish Mathematician in Germany.* New York: Springer, 2016.

Frege, Gottlob. *Grundgesetze der Arithmetik.* Jena, Germany: Hermann Pohle, 1893.

Gierer, Alfred. "Gödel Meets Carnap: A Prototypical Discourse on Science and Religion." *Zygon* 32, no. 2 (June 1997): 207–17.

Gödel, Rudolf. "History of the Gödel Family." In *Gödel Remembered: Salzburg, 10–12 July 1983*, edited by Paul Weingartner and Leopold Schmetterer. Naples: Bibliopolis, 1987.

———. "Skizze zu einer Chronik der Familie Gödel." In *Kurt Gödel: Wahrheit & Beweisbarkeit*, vol. 1: *Dokumente und historische Analysen*, edited by Eckehart Köhler et al. Vienna: öbv & hpt, 2002.

Goldstein, Rebecca. *Incompleteness: The Proof and Paradox of Kurt Gödel.* New York: Norton, 2005.

Goldstine, Herman. Department of Mathematics Oral History Project. The Princeton Mathematics Community in the 1930s. Transcript No. 15. Seeley G. Mudd Manuscript Library, Princeton University, Princeton, N.J.

Graf, Max. *Legend of a Musical City.* New York: Philosophical Library, 1945.

Graham, Elyse. "Adventures in Fine Hall." *Princeton Alumni Weekly*, January 10, 2018.

Guerra-Pujol, F. E. "Gödel's Loophole." *Capital University Law Review* 41, no. 3 (Summer 2013): 637–74.

Hahn, Hans, et al. "Diskussion zur Grundlegung der Mathematik." *Erkenntnis* 2 (1931): 135–51. English translation in "Discussion on the Foundation of Mathematics," *History and Philosophy of Logic* 5, no. 1 (January 1984): 111–29.

Halperin, Israel. Department of Mathematics Oral History Project. The Princeton Mathematics Community in the 1930s. Transcript No. 18. Seeley G. Mudd Manuscript Library, Princeton University, Princeton, N.J.

Hamann, Brigitte. *Hitler's Vienna.* New York: Oxford University Press, 1999.

Heindl, Waltraud. *Gehorsame Rebellen: Bürokratie und Beamte in Österreich.* Vienna: Böhlau, 1991.

Helmer, Olaf. "Perelman versus Gödel." *Mind* 46, no. 181 (January 1937): 58–60.

Hilbert, David. "Mathematische Probleme." *Nachrichten von der Gesellschaft der Wissenschaften zu Göttingen, Mathematisch-physikalisch* (1900): 253–97. English translation, "Mathematical Problems," *Bulletin of the American Mathematical Society* 8, no. 10 (July 1902): 437–79.

———. "On the Infinite." 1925. Reprinted in *From Frege to Gödel: A Source Book in Mathematical Logic, 1879–1932*, edited by Jean van Heijenoort. Cambridge: Harvard University Press, 1967.

Hoffmann, Dirk W. *Die Gödel'schen Unvollständigkeitssätze: Eine geführte Reise durch Kurt Gödels historischen Beweis.* 2nd ed. Berlin: Springer, 2017.

———. *Grenzen der Mathematik: Eine Reise durch die Kerngebiete der mathematischen Logik.* 3rd ed. Berlin: Springer, 2018.

Hofmannsthal, Hugo von. "Die österreichische Idee." 1917. Reprinted in *Prosa III*, edited by Herbert Steiner. Frankfurt: S. Fischer, 1951.

Holton, Gerald, and Yehuda Elkana, eds. *Albert Einstein: Historical and Cultural Perspectives.* Princeton: Princeton University Press, 1982.

Huber-Dyson, Verena. "Gödel and the Nature of Mathematical Truth II." Interview on Edge.org, July 26, 2005.

Institute for Advanced Study. Archives. Files on Kurt Gödel. Director's Office: Faculty Files, Princeton, N.J.

———. Director's Office: Visa-Immigration Files.

———. School of Mathematics Records: Faculty Files.

Institute for Advanced Study. History Working Group. *A Refuge for Scholars.* Princeton, N.J., 2017.

"Interview with Martin Davis." *Notices of the American Mathematical Society 55*, no. 5 (May 2008): 560–71.

Janik, Allan, and Stephen Toulmin. *Wittgenstein's Vienna.* 1973. Chicago: Irving R. Dee, 1996.

Johnson, Lonnie. *Introducing Austria.* Riverside, Calif.: Ariadne Press, 1989.

Johnston, William M. *The Austrian Mind: An Intellectual and Social History, 1848–1938.* 1972. Berkeley: University of California Press, 1983.

Jones, Kingsley. "Insulin Coma Therapy in Schizophrenia." *Journal of the Royal Society of Medicine 93*, no. 3 (March 2000): 147–49.

Jones, Landon Y., Jr. "Bad Days on Mount Olympus." *The Atlantic*, February 1974, 37–53.

Judson, Peter M. *The Habsburg Empire: A New History.* Cambridge: Harvard University Press, 2018.

Kleene, Stephen C. "Kurt Gödel, 1906–1978: A Biographical Memoir." Washington, D.C.: National Academy of Sciences, 1987.

Klepetar, Harry E. "The Chances of Communism in China." *American Scholar 19*, no. 3 (Summer 1950): 301–8.

Köhler, Eckehart. "The Philosophy of Misdeed." Unpublished manuscript, 1968.

Köhler, Eckehart, et al., eds. *Kurt Gödel: Wahrheit & Beweisbarkeit.* Vol. 1, *Dokumente und historische Analysen.* Vienna: öbv & hpt, 2002.

Kraus, Karl. "Franz Ferdinand und die Talente." *Die Frackel*, no. 400, July 10, 1914.

―――. *The Last Days of Mankind: A Tragedy in Five Acts*. 1920. Translated by Patrick Healy. Amsterdam: November Editions, 2016.

Kreisel, Georg. "Gödel's Excursions into Intuitionist Logic." In *Gödel Remembered: Salzburg, 10–12 July 1983*, edited by Paul Weingartner and Leopold Schmetterer. Naples: Bibliopolis, 1987.

―――. "Kurt Gödel." *Biographical Memoirs of Fellows of the Royal Society* 26 (1980): 158–224.

Kuh, Anton. "'Central' und 'Herrenhof'." In *Der unsterbliche Österreicher*. Munich: Knorr & Hirth, 1931.

List, Rudolf. *Brünn, ein deutsches Bollwerk*. St. Pölten, Austria: St. Pöltner Zeitungs-Verlags, 1942.

Mancosu, Paolo. "Between Vienna and Berlin: The Immediate Reception of Gödel's Incompleteness Theorems." *History and Philosophy of Logic* 20, no. 1 (January 1999): 33–45.

Martin, Donald A. "Gödel's Conceptual Realism." *Bulletin of Symbolic Logic* 11, no. 2 (June 2005): 207–24.

May, Arthur James. *The Hapsburg Monarchy, 1867–1914*. New York: Norton, 1968.

Maynard, W. Barksdale. "Princeton in the Confederacy's Service." *Princeton Alumni Weekly*, March 23, 2011.

Menger, Karl. *Ergebnisse eines Mathematisches Kolloquiums*. Edited by Egbert Dierker and Karl Sigmund. Vienna: Springer, 1998.

―――. "Introduction." In *Empiricism, Logic and Mathematics: Philosophical Papers* by Hans Hahn, edited by Brian McGuinness. Dordrecht, Netherlands: D. Reidel, 1980.

―――. *Reminiscences of the Vienna Circle and the Mathematical Colloquium*. Edited by Louise Golland, Brian McGuinness, and Abe Sklar. Dordrecht, Netherlands: Kluwer, 1994.

Monk, Ray. *Bertrand Russell: The Spirit of Solitude, 1872–1921*. New York: Free Press, 1996.

―――. *Ludwig Wittgenstein: The Duty of Genius*. 1990. New York: Penguin, 1991.

Morgenstern, Oskar. "Abraham Wald, 1902–1950." *Econometrica* 19, no. 4 (October 1951): 361–67.

―――. Diaries. Oskar Morgenstern Papers. Rare Book and Manuscript Library, Duke University, Durham, N.C. Digital publication on the Web, "Oskar Morgenstern Tagebuchedition," University of Graz, Austria.

Musil, Robert. *Der Mann ohne Eigenschaften*. 1930. Hamburg: Rowohlt, 1952.

————. "Der mathematische Mensch." 1913. In *Gesammelte Werke*, vol. 2. Hamburg: Rowohlt, 1978.

Nagel, Ernest. "Impressions and Appraisals of Analytic Philosophy in Europe. II." *Journal of Philosophy* 33, no. 2 (January 16, 1936): 29–53.

Nagel, Ernest, and James R. Newman. *Gödel's Proof.* 1958. Rev. ed. New York: New York University Press, 2001.

National Archives and Records Administration. Membership Applications to the NS-Frauenschaft/Deutsches Frauenwerk. Microfilm Publication A3344. National Archives, College Park, Md.

————. Nazi Party Applications by Austrians (1938–39). Microfilm Publication A3359.

Pais, Abraham. *"Subtle Is the Lord . . .": The Science and the Life of Albert Einstein.* Oxford: Oxford University Press, 1982.

Patterson, K. David, and Gerald F. Pyle. "The Geography and Mortality of the 1918 Influenza Pandemic." *Bulletin of the History of Medicine* 65, no. 1 (Spring 1991): 4–21.

Pauley, Bruce F. *From Prejudice to Persecution: A History of Austrian Anti-Semitism.* Chapel Hill: University of North Carolina Press, 2000.

————. *Hitler and the Forgotten Nazis: A History of Austrian National Socialism.* Chapel Hill: University of North Carolina Press, 1981.

Perelman, Chaïm. "Les Paradoxes de la Logique." *Mind* 45, no. 178 (April 1936): 204–8.

Pfefferle, Roman, and Hans Pfefferle. *Glimpflich entnazifiziert: Die Professorenschaft der Universität Wien von 1944 in den Nachkriegsjahren.* Göttingen: V&R unipress, 2014.

Post, Emil L. "Finite Combinatory Processes—Formulation 1." *Journal of Symbolic Logic* 1, no. 3 (September 1936): 103–5.

Raatikainen, Panu. "On the Philosophical Relevance of Gödel's Incompleteness Theorems." *Revue Internationale de Philosophie* 59, no. 4 (October 2005): 513–34.

Regis, Ed. *Who Got Einstein's Office? Eccentricity and Genius at the Institute for Advanced Study.* Reading, Mass.: Addison-Wesley, 1987.

Reid, Constance. *Hilbert.* 1970. New York: Springer, 1996.

Roth, Joseph. *The Emperor's Tomb.* 1938. Translated by John Hoare. New York: Overlook Press, 2002.

————. *The Radetzky March.* 1932. Translated by Joachim Neugroschel. New York: Overlook Press, 1995.

Rothkirchen, Livia. *The Jews of Bohemia and Moravia: Facing the Holocaust.* Lincoln: University of Nebraska Press, 2012.

Rothschild, Joseph. *East Central Europe Between the Two World Wars*. Seattle: University of Washington Press, 1974.

Rudin, Walter. *The Way I Remember It*. History of Mathematics, vol. 12. Providence, R.I.: American Mathematical Society and London Mathematical Society, 1996.

Russell, Bertrand. *The Autobiography of Bertrand Russell, 1987–1914*. Boston: Little, Brown, 1967.

———. *The Autobiography of Bertrand Russell, 1914–1944*. Boston: Little, Brown, 1968.

———. *A History of Western Philosophy*. 1945. New York: Simon and Schuster, 1967.

———. *Introduction to Mathematical Philosophy*. London: George Allen and Unwin, 1919.

———. *My Philosophical Development*. 1959. London: Routledge, 1993.

Sacks, Gerald. "Reflections on Gödel." 3rd annual Thomas and Yvonne Williams Lecture for the Advancement of Logic and Philosophy, University of Pennsylvania, 2007.

Schewe, Phillip F. *Maverick Genius: The Pioneering Odyssey of Freeman Dyson*. New York: St. Martin's, 2013.

Schlick, Moritz. Einstein-Schlick Correspondence. ECHO, Max Planck Institute for the History of Science. echo.mpiwg-berlin.mpg.de.

———. Nachlass. Papers of the Vienna Circle Movement. Noord-Hollands Archief, Haarlem, Netherlands.

Schober-Bendixen, Susanne. *Die Tuch-Redlichs: Geschichte einer jüdischen Fabrikantenfamilie*. Vienna: Amalthea, 2018.

Schorske, Carl. *Fin-de-siècle Vienna: Politics and Culture*. New York: Vintage, 1981.

Segal, Sanford L. "Mathematics and German Politics: The National Socialist Experience." *Historia Mathematica* 13, no. 2 (May 1986): 118–35.

Segel, Harold B., ed. *The Vienna Coffeehouse Wits, 1890–1938*. West Lafayette, Ind.: Purdue University Press, 1993.

Selberg, Atle. Oral History Project. Institute for Advanced Study Archives, Princeton, N.J.

———. Papers. Institute for Advanced Study Archives, Princeton, N.J.

Siegmund-Schultze, Reinhard. *Mathematicians Fleeing from Nazi Germany: Individual Fates and Global Impact*. Princeton: Princeton University Press, 2009.

———. " 'Mathematics Knows No Races': A Political Speech that David Hilbert Planned for the ICM in Bologna in 1928." *Mathematical Intelligencer* 38, no. 1 (March 2016): 56–66.

Sigmund, Karl. "Dozent Gödel Will Not Lecture." In Mathias Baaz et al., eds., *Kurt Gödel and the Foundations of Mathematics*. Cambridge: Cambridge University Press, 2011.

———. *Exact Thinking in Demented Times: The Vienna Circle and the Epic Quest for the Foundations of Science*. New York: Basic Books, 2017.

———. "A Philosopher's Mathematician: Hans Hahn and the Vienna Circle." *Mathematical Intelligencer* 17, no. 4 (Fall 1995): 16–29.

Sigmund, Karl, John Dawson, and Kurt Mühlberger. *Kurt Gödel: Das Album/ The Album*. Wiesbaden: Vieweg, 2006.

Simotta, Daphne-Ariane. "Marriage and Divorce Regulation and Recognition in Austria." *Family Law Quarterly* 29, no. 3 (Fall 1995): 525–40.

Slezak, Leo. *Rückfall*. Stuttgart: Rowohlt, 1940.

Smith, Alice Kimball. "The Elusive Dr. Szilard." *Harper's*, August 1960.

Snapper, Ernst. "The Three Crises in Mathematics: Logicism, Intuitionism and Formalism." *Mathematics Magazine* 52, no. 4 (September 1979): 207–16.

Sokal, Alan, and Jean Bricmont. *Fashionable Nonsense: Postmodern Intellectuals' Abuse of Science*. New York: Picador, 1998.

Spiel, Hilde. *Vienna's Golden Autumn, 1866 to 1938*. New York: Weidenfeld and Nicolson, 1987.

Stadler, Friedrich. *The Vienna Circle: Studies in the Origins, Development, and Influence of Logical Empiricism*. Rev. ed. Heidelberg: Springer, 2015.

Stadler, Friedrich, and Peter Weibel, eds. *Vertreibung der Vernunft: The Cultural Exodus from Austria*. 2nd ed. Vienna: Springer, 1995.

Stern, Beatrice M. "A History of the Institute for Advanced Study, 1930–1950." Unpublished manuscript. Institute for Advanced Study Archives, Princeton, N.J.

Taussky-Todd, Olga. Oral History Project. California Institute of Technology Archives, Pasadena, Calif., 1979–80.

———. "Remembrances of Kurt Gödel." In *Gödel Remembered: Salzburg, 10–12 July 1983*, edited by Paul Weingartner and Leopold Schmetterer. Naples: Bibliopolis, 1987.

Thomas, Dorothy Morgenstern. Collection. Institute for Advanced Study Archives, Princeton, N.J.

———. Oral History Project. Institute for Advanced Study Archives, Princeton, N.J.

Topp, Leslie. "An Architecture for Modern Nerves: Josef Hoffmann's Purkersdorf Sanatorium." *Journal of the Society of Architectural Historians* 56, no. 4 (December 1997): 414–37.

Tucker, Albert. Department of Mathematics Oral History Project. The

Princeton Mathematics Community in the 1930s. Transcript No. 30. Seeley G. Mudd Manuscript Library, Princeton University, Princeton, N.J.

Turing, A. M. "On Computable Numbers." *Proceedings of the London Mathematical Society* 42 ser. 2 (1937): 230–65.

Ulam, S. M. *Adventures of a Mathematician.* 1976. Berkeley: University of California Press, 1991.

University of Vienna. *650 Plus—History of the University of Vienna.* geschichte.univie.ac.at.

Van Atten, Mark. "Gödel and Brouwer: Two Rivalling Brothers." In *Essays on Gödel's Reception of Leibniz, Husserl, and Brouwer.* New York: Springer, 2014.

Van Heijenort, Jean, ed. *From Frege to Gödel: A Source Book in Mathematical Logic, 1879–1931.* Cambridge: Harvard University Press, 1967.

Veblen, Oswald. Papers. Manuscript Division, Library of Congress, Washington, D.C.

Vinnikov, Victor. "We Shall Know: Hilbert's Apology." *Mathematical Intelligencer* 21, no. 1 (March 1999): 42–46.

Von Kármán, Theodore. *The Wind and Beyond.* Boston: Little, Brown, 1967.

Von Neumann, John. *The Computer and the Brain.* New Haven: Yale University Press, 1958.

Von Plato, Jan. "In Search of the Sources of Incompleteness." *Proceedings of the International Congress of Mathematicians,* Rio de Janeiro, 2018, 3:4043–60.

Walker, Mark. "National Socialism and German Physics." *Journal of Contemporary History* 24, no. 1 (January 1989): 63–89.

Wang, Hao. *A Logical Journey: From Gödel to Philosophy.* Cambridge, Mass.: MIT Press, 1996.

———. *Reflections on Kurt Gödel.* Cambridge, Mass.: MIT Press, 1987.

———. "Some Facts about Kurt Gödel." *Journal of Symbolic Logic* 46, no. 3 (September 1981): 653–59.

Wasserman, Janek. *Black Vienna: The Radical Right in the Red City, 1918–1938.* Ithaca, N.Y.: Cornell University Press, 2014.

Watterson, Kathryn. *I Hear My People Singing: Voices of African American Princeton.* Princeton: Princeton University Press, 2017.

Waugh, Alexander. *The House of Wittgenstein: A Family at War.* New York: Doubleday, 2008.

Weber, Max. "Science as a Vocation." 1917. In *Max Weber: Essays in Sociology,* translated and edited by H. H. Gerth and C. Wright Mills. New York: Oxford University Press, 1946.

Weil, André. Director's Office: Faculty Files. Institute for Advanced Study Archives, Princeton, N.J.

Weingartner, Paul, and Leopold Schmetterer, eds. *Gödel Remembered: Salzburg, 10–12 July 1983*. Naples: Bibliopolis, 1987.

Weyl, Hermann. "David Hilbert and His Mathematical Work." *Bulletin of the American Mathematical Society* 50, no. 9 (September 1944): 612–54.

———. "Mathematics and Logic." *American Mathematical Monthly* 53, no. 1 (January 1946): 2–13.

Whitehead, Alfred North, and Bertrand Russell. *Principia Mathematica*. 3 vols. Cambridge: Cambridge University Press, 1910–13.

Whitman, Marina von Neumann. *The Martian's Daughter: A Memoir*. Ann Arbor: University of Michgan Press, 2012.

Wigderson, Avi. "The Gödel Phenomenon in Mathematics: A Modern View." In *Kurt Gödel and the Foundations of Mathematics*, edited by Mathias Baaz et al. Cambridge: Cambridge University Press, 2011.

Winkler, Wilhelm. "The Population of the Austrian Republic." *Annals of the American Academy of Political and Social Science* 98, supp.: Present Day Social and Industrial Conditions in Austria (November 1921): 1–6.

Woolf, Harry, ed. *Some Strangeness in the Proportion: A Centennial Symposium to Celebrate the Achievements of Albert Einstein*. Reading, Mass.: Addison-Wesley, 1980.

Zahra, Tara. *Kidnapped Souls: National Indifference and the Battle for Children in the Bohemian Lands, 1900–1948*. Ithaca, N.Y.: Cornell University Press, 2011.

Zimmermann, Volker. *Die Sudetendeutschen im NS-Staat: Politik und Stimmung der Bevölkerung im Reichsgau Sudetenland*. Essen: Klartext, 1999.

Zuckmayer, Carl. *A Part of Myself*. 1966. Translated by Richard and Clara Winston. New York: Harcourt Brace Jovanovich, 1970.

Zweig, Stefan. *The World of Yesterday*. 1942. Rev. ed., translated by Anthea Bell. Lincoln: University of Nebraska Press, 2009.

ILLUSTRATION CREDITS

KGP: Kurt Gödel Papers, the Shelby White and Leon Levy Archives Center, Institute for Advanced Study, Princeton, N.J., on deposit at Princeton University Library

ÖNB: Austrian National Library, Photo Archive

IAS: Shelby White and Leon Levy Archives Center, Institute for Advanced Study, Princeton, N.J.

91   R. G. Lubben Papers, Archives of American Mathematics, e_math_00018, The Dolph Briscoe Center for American History, University of Texas at Austin

107 (top) KGP, 14b/17, 110064

107 (bottom) IAS, Unsorted People letter box 1, Adele Gödel

126 KGP, 7b/13, 040021

138 KGP, 13b/44, 090722

140 (top) KGP, 14b/17, 110026

140 (bottom) ÖNB, 207.229B

151 IAS, OV SM3, photograph by Wilhelm Blaschke

157 Alfred Eisenstaedt LIFE Picture Collection/Getty Images

161 ÖNB, OEGZ H 780B

163 Courtesy Bancroft Library, University of California, Berkeley

167 KGP, 13b/45, 090755

175 KGP, 14a/0, 110153

182 ÖNB, OEGZ H 4531B

186 KGP, 6c/81, 030114

191 ÖNB, 99.113B

197 ÖNB, OEGZ S 283/27

198 KGP, 13b/18, 090374

199 KGP, 14a/1, 110206

205 KGP, 13a/8; IAS, Director's Office: Faculty Files, Kurt Gödel, Pre-1953.

206 KGP, 14b/17, 110067

209 IAS, BP 03, photograph by Rose and Son, Princeton, N.J.

215 Courtesy George Dyson, photograph by Verena Huber-Dyson

217 IAS, EB 069, photograph by Oskar Morgenstern

219 IAS, SM Goe 02, photograph by Dorothy Morgenstern Thomas

252 KGP, 14b/17, 110071

253 KGP, 14b/17, 110033

254 KGP, 14b/17, 110133

262 KGP, 14b/17, 110121

270 IAS, Postcard

271 Courtesy Library of Congress, ppmsca 56705, photograph by John T. Bledsoe

273 Courtesy IAS, SM Goe 03, photograph by A. G. Wightman

INDEX

Page references to maps and illustrations are in **bold** type.